STRUCTURE AND BONDING

Volume 22

Editors: J. D. Dunitz, Zürich
P. Hemmerich, Konstanz · R. H. Holm, Cambridge
J. A. Ibers, Evanston · C. K. Jørgensen, Genève
J. B. Neilands, Berkeley · D. Reinen, Marburg
R. J. P. Williams, Oxford

With 36 Figures

Springer-Verlag
Berlin Heidelberg GmbH 1975

ISBN 978-3-662-15534-9   ISBN 978-3-540-37563-0 (eBook)
DOI 10.1007/978-3-540-37563-0

Library of Congress Catalog Card Number 67-11280

Contents

Manuscripts will be accepted by the editors:

Professor Dr. *Jack D. Dunitz*	Laboratorium für Organische Chemie der Eidgenössischen Hochschule CH-8006 Zürich, Universitätsstraße 6/8
Professor Dr. *Peter Hemmerich*	Universität Konstanz, Fachbereich Biologie D-7750 Konstanz, Postfach 733
Professor *Richard H. Holm*	Department of Chemistry, Massachusetts Institute of Technology Cambridge, Massachusetts 02139/U.S.A.
Professor *James A. Ibers*	Department of Chemistry, Northwestern University Evanston, Illinois 60201/U.S.A.
Professor Dr. *C. Klixbüll Jørgensen*	51, Route de Frontenex, CH-1207 Genève
Professor *Joe B. Neilands*	University of California, Biochemistry Department Berkeley, California 94720/U.S.A.
Professor Dr. *Dirk Reinen*	Fachbereich Chemie der Universität Marburg D-3550 Marburg, Gutenbergstraße 18
Professor *Robert Joseph P. Williams*	Wadham College, Inorganic Chemistry Laboratory Oxford OX1 3QR/Great Britain

SPRINGER-VERLAG

D-6900 Heidelberg 1
P. O. Box 105280
Telephone (06221) 487·1
Telex 04-61723

D-1000 Berlin 33
Heidelberger Platz 3
Telephone (030) 822001
Telex 01-83319

SPRINGER-VERLAG
NEW YORK INC.

175, Fifth Avenue
New York, N. Y. 10010
Telephone 673-2660

The Lanthanide Ions as Structural Probes in Biological and Model Systems

Evert Nieboer

Department of Chemistry, Laurentian University, Sudbury, Ontario P3E 2C6, Canada

Table of Contents

Introduction

The substitution of metal ions in biochemical macromolecules by others of similar chemical reactivity but possessing improved physical properties has proven to be a versatile research tool. Considerable structural information has been obtained by exchanging manganese(II) for magnesium [e.g. (1)], cobalt(II) for zinc [e.g. (2)] and thallium(I) for potassium (3). *Vallee* and *Williams* (4) in 1968 suggested the potential of rare earth cations as reporter sites in biological systems requiring calcium. Shortly afterwards, *Birnbaum et al.* (5) reported that binding of Nd(III) to bovine serum albumin could be detected by difference spectrophotometry, and *Williams* and co-workers (6, 7) demonstrated that lanthanide cations were excellent NMR probes of their immediate environment in enzymes. The latter group were able to estimate absolute distances from the Gd(III) reporter site in lysozyme to protons on the bound substrate, β-methyl-N-acetylglucosamine, from isotropic broadening of the corresponding substrate proton signals. Somewhat earlier, *Hinckley* (8) had reported the influence of the dipyridine adduct of Eu(thd)$_3$[1] on the PMR spectrum of cholesterol in carbon tetrachloride. The subsequent exponential development of the principles, methodology and application of lanthanide β-diketonates as NMR shift reagents in Organic Chemistry has been extensively documented in recent reviews (9—13), and this topic will not be examined in this context in the present article. This review will be addressed to an assessment of the methods employed and their success in providing information about the immediate and distant environment of the trivalent lanthanide ions when incorporated in model systems, or when complexed with low-molecular weight molecules and macromolecules of biological interest. This application of lanthanide ions as structural probes is possible because of favourable physicochemical properties. Consequently, a brief survey of relevant lanthanide coordination chemistry is also provided.

I. Coordination Chemistry of the Tripositive Lanthanide Ions

i) Nature of Complexes

A compilation of typical co-ordination polyhedra known for the tripositive rare earths (RE) is given in Table 1. For comparison, crystallographic data for calcium is also included. The lanthanide(III) ions, like the alkaline earth ions, have been classified as class a (57, 58) or "hard" (59, 60) acceptors. Consequently, their interactions may be considered as largely ionic, since the consensus is that the gross features of class a and b behaviour are satisfactorily explained by the relative importance of ionic and covalent bonding (57—62a). It is expected, and observed, that the RE ions prefer donor atoms with the preference $O > N > S$ and $F > Cl$. In fact, the lanthanide-halogen stretching force constant decreases from fluorine to iodine (62b). Complexes with pure nitrogen or sulfur donors are readily hydrolyzed (56, 63) and are not stable in aqueous solutions. It is not surprising therefore that oxygen is the predominant donor atom in the structures described in Table 1.

---

[1] Tris-(2,2,6,6-tetramethylheptane-3,5-dionato) europium(III).

Table 1. Crystallographic data for complexes of the trivalent lanthanides and calcium

Complex[a]	Metal ion	CN[b]	Polyhedron[c]	Remarks	Ref.
[Ln(EDTAH) (H$_2$O)$_4$] · 3 H$_2$O	La	10	bcdod*		(14, 15)
Ln(dipyr)$_2$(NO$_3$)$_3$	La	10	bcdod*	nitrates are bidentate	(16–18)
Na,K[M(EDTA) (H$_2$O)$_3$] · XH$_2$O	La, Tb, (Ca)	9	mc-sqantipr	Ln=La, Nd, Gd, Tb, Er are isomorphous	(15, 19, 20)
[Ln(OH$_2$)$_9$](BrO$_3$)$_3$	Nd	9	tctp	r_{eq}(Nd–O) ≃ r_{ax}(Nd–O)	(15)
[Ln(OH$_2$)$_9$] (C$_2$H$_5$SO$_4$)$_3$	Er, Pr	9	tctp	r_{eq} > r_{ax} by ~0.15 Å	(15)
[M(H$_2$O)$_6$]Cl$_2$	Sr, Ba	9	tctp	r_{eq} > r_{ax}	(15, 21)
Ln(HOCH$_2$COO)$_3$	Gd, La	9	tctp*	Ln=La–Tb, orthor. form, are isomorphous; alcohol O bound equatorially	(22, 23, 24)
Na$_3$[Ln(O$_2$CCH$_2$OCH$_2$CO$_2$)$_3$] · 6H$_2$O, 2NaClO$_4$	Nd, Yb	9	tctp*	Ln=Ce–Lu isomorphous; ether O bound equatorially	(25, 26)
Na$_3$[Ln(terpy)$_3$](ClO$_4$)$_3$	Eu	9	tctp*	9 N-atoms bound to metal	(18, 27)
Na$_3$[Ln(dp)$_3$] · XH$_2$O	Nd, Yb	9	tctp*	pyridine-N bound equatorially	(28)
Ln(thd)$_3$X$_2$ X = 4-pic X = pyr	Ho Eu	8	sqantipr	X is on different square faces	(29, 30)
Ln(acac)$_3$(H$_2$O)$_2$	La	8	sqantipr	waters on same square face	(31)
[Ln(H$_2$O)$_6$Cl$_2$]Cl	Eu, Gd	8	sqantipr*	Ln=Nd, Sm, Er, isomorphous	(15)
[M(H$_2$O)$_8$]O$_2$	Ca, Sr, Ba	8	sqantipr*		(15, 32)
Ln(PO$_4$)	Ce, La, Nd	8	dod	bridging phosphate oxygens	(15)
M(H$_x$PO$_4$)$_{1,2}$(H$_2$O)$_y$	Ca	8,7	dod(?), pentbipyr	bridging phosphate oxygens; x = 1,2; y = 0–2	(33, 34)
[M(O$_2$CCH$_2$OCH$_2$CO$_2$) (H$_2$O)$_5$] · H$_2$O	Ca	8	—	r(ether) ≃ r(COO$^-$)	(35)
[M(sugar) (H$_2$O)$_x$]Br$_{1,2}$ · yH$_2$O	Ca	8	sqantipr*	sugars = lactose, galactose, etc. lactobionic acid, etc.	(36, 37)

3

Table 1 (continued)

Complex[a]	Metal ion	CN[b]	Polyhedron[c]	Remarks	Ref.
[Ln₂(fod)₆(H₂O)]·H₂O	Pr	8	dod and bctp	bridging water	(38)
Ln₂(thd)₆	Pr	7	mtcp*	Ln=La–Dy, monoclinic, isomorphous	(39)
Ln(thd)₃X X = H₂O	Dy	7	mctp*	X not the cap	(40–42)
X = 3–CH₃-pyr	Lu	7	mctp*	X not the cap	
X = DMSO	Eu	7	pentbipyr*	DMSO in axial positions	
M V₃O₇	Ca	7	mctp	bridging oxygens	(43)
[M(trigly)(H₂O)₂]Cl₂·H₂O	Ca	7	pentbipyr	donors: carbonyl, carboxylate, water	(44)
[M(thymidylite)(H₂O)₃]·3H₂O	Ca	7	pentbipyr	water in axial positions	(45)
M(⁺NH—CH₂CO₂)₃(H₂O)₂	Ca	7	pentbipyr	seven O-atoms form polyhedron	(46)
Ln(thd)₃	Er	6	tp	Ln=Tb–Lu, orthorhombic, isomorphous	(47)
MO, M(OH)₂	Ca	6	oct		(21, 48)
Paravalbumin	Ca	6	oct*	donors: carbonyl, carboxylate alcohol	(49)
Concanavalin A	Ca	6	oct*	donors: carbonyl, carboxylate, water	(50)
Staphylococcal nuclease	Ca	6?	oct*?	polyhedron not rigorously defined	(51, 52)
Thermolysin	Ca, Ln	Ca (6,7); Ln (≥8)	irregular	donors: carbonyl, carboxylate, alcohol, water	(53, 54)
[Ln(NCS)₆](⁺N—R₄)₃	Er	6	oct	bonding through N of NCS⁻	(55, 56)

a) Abbreviations: Ln, Lanthanide; EDTA(H), ethylenediame-N,N,N',N'-tetra-acetic acid; dipyr, 2,2'-dipyridyl; terpy, terpyridyl; dp, pyridine-2,6-dicarboxylic acid (dipicolinate); thd, 2,2,6,6-tetramethyl-heptane-3,5-dione; acac, acetylacetonate; 4-pic, 4-methylpyridine; pyr, pyridine; fod, 1,1,1,2,2,3,3,-heptafluoro-7,7-dimethyl-4,6-octanedione; trigly, glycylglycylglycine; N–R₄, tetrabutylammonium ion.

b) CN, coordination number.

c) Abbreviations: bcdod, bicapped dodecahedron; mc-sqantipr, monocapped square antiprism; tctp, tricapped trigonal prism; sqantipr, square antiprism; c pentbipyr, capped pentagonal bipyramid; pentbipyr, pentagonal bipyramid; mctp, monocapped trigonal prism; tp, trigonal prism; oct, octahedron.

For a detailed description of these co-ordination polyhedra see the review by *Muetterties* and *Wright* (15).

The high degree of ionicity in these complexes is further confirmed by the observed patterns of bond distances. Especially noteworthy is that Ln—N bonds are generally longer than Ln—O bonds, and in some cases, by as much as 0.30 Å.

The co-ordination numbers reported for the RE ions range from six to ten, although a maximum value of twelve is known (56). Co-ordination numbers of eight and nine would appear to be most common, while Ca^{+2} exhibits numbers of six, seven and eight. This tendency of the RE ions to have larger numbers of donor atoms in their primary co-ordination sphere could be crucial in isomorphous substitution experiments, especially in calcium dependent systems. We will return to this comment in a later section. Preference for a specific geometry may also be significant in this regard. For example, the most common seven co-ordinate structure observed for Ca^{+2} is the pentagonal bipyramid, while that for the lanthanide ions would appear to be the monocapped trigonal prism (41, 47).

One additional observation worthy of comment may be derived from the structural data in Table 1. In tricapped trigonal prismatic structures of complexes with ligands composed of both charged and neutral donor groups, the latter occupy the equatorial or cap positions. The bonds to these donor atoms tend to be marginally longer. Thus the alcohol oxygen of glycolate, the ether oxygen of diglycolate and the nitrogen atom of dipicolinate cap the trigonal prism. It must be concluded that all nine sites of this structure are not equivalent in these complexes. Neither are Ln—O separations in other structures necessarily equivalent [e.g. (40)].

ii) Complex Formation in Solution

The shapes of curves in plots of complex stability *versus* atomic number (Fig. 1) are varied, and would appear to possess some diagnostic potential. One of the most spectacular dependencies on position in the RE series is exhibited by K_1 values[2]) for acetate and diglycolate. A reduction in complex stability occurs around the middle of the series in both cases. In contrast, the curves for log K_1 of EDTA and acetylacetonate reveal a systematic increase in negative free energy across the series, although the rate of increase in both halves of the series is considerably different. All the plots in Fig. 1 show a rather drastic change in slope or contour around the middle of the series.

This feature has been designated as the "gadolinium break", or when examined in greater detail, the "tetrad" effect (68). Much has been said in explanation of these trends. Of these, the suggestion of a change in the primary hydration number of the aquo ions between neodymium and terbium is the most popular interpretation (69). Others such as ligand-field effects and steric factors have also been postulated (56, 70). Ligand-field stabilization was shown to be at least one order of magnitude too weak to be consistent with the experimental data, although the possibility of ligand-field stabilization involving the 5d shell or other empty orbitals of the central cation should not be completely excluded (68). Small decreases in the interelectronic repulsion parameters (nephelauxetic

---

[2]) $M + L = ML$ $K_1 = [ML]/[M][L]$.

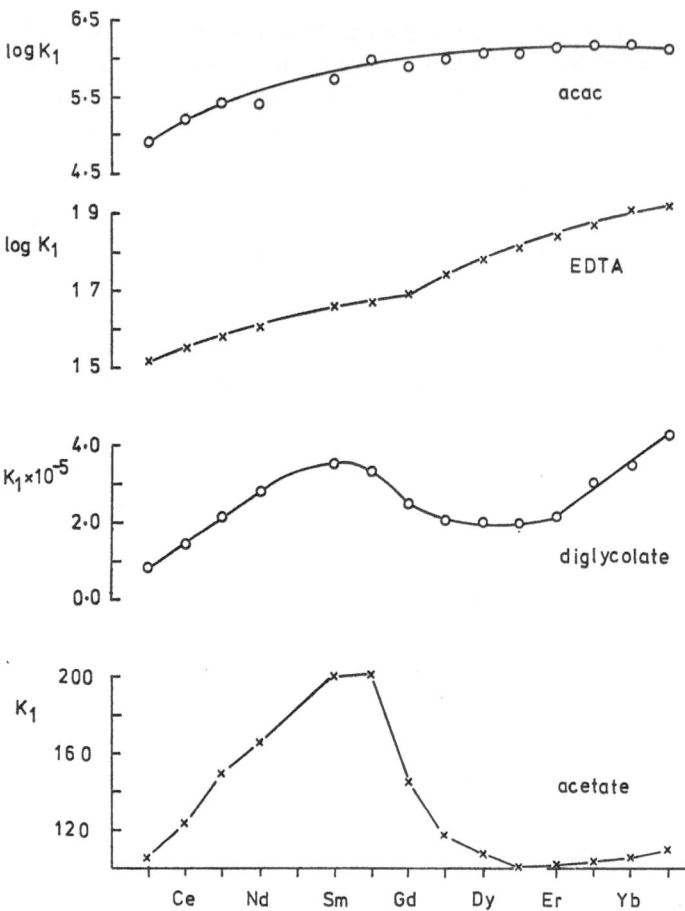

Fig. 1. Stability of complexes with acetate (*64*), diglycolate (*65*) EDTA (*66*) and acetylacetonate (*67*) as a function of atomic number

effect) on complex formation could, in principle, be responsible for the gadolinium break and other smaller irregularities of the "tetrad" effect (*68*). A change in the degree of covalency relative to the aquo complex is intrinsic in this explanation, and this is difficult to verify. The sudden drop in the third ionization potential (Fig. 2) between europium and gadolinium and in other electronic properties (*68*) should reflect the relative tendencies of the RE ions to form covalent bonds. A plot of the class b index $(\sum_{n=1}^{3} I_n) r_{cat}$ *versus* the cationic radius r_{cat} (*58*) divides the trivalent ions into two groups. Ions in the series La—Eu show distinct enhanced class b character [*i.e.*, relative magnitudes of $(\sum_{n=1}^{3} I_n) r_{cat}$ and r_{cat} are both large] compared to the ions in the series Gd—Lu. On this basis, covalency is expected to be most important in the first half of the series; but of course, nevertheless of secondary importance for reactions in aqueous media.

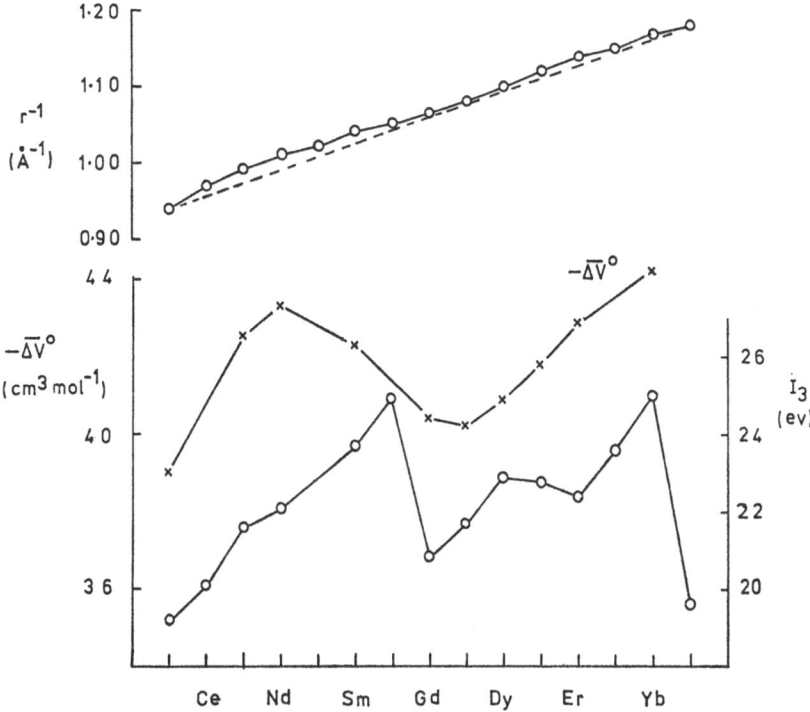

Fig. 2. Reciprocal cationic radii (63), partial molal volumes (72), and the third ionization potential (73) as a function of atomic number

It will be shown in a subsequent section that the variation of complex stability across the RE series may assist in determining the nature of a metal binding site. A closer examination of this phenomenon is thus warranted. *Phillips* and *Williams* (71) some time ago postulated that in the absence of covalency, energy changes across the RE series should resemble the plot of reciprocal cationic radius depicted in Fig. 2. The degree of curvature and the gradient in each half of the series was predicted to vary with the nature of the chemical partner of the lanthanides [e.g. free energies of formation of $Ln^{+3}(aq)$ and $Ln(OH)_3$]. Energy changes accompanying chemical reactions would then be determined by differences between such plots, which conceivably could resemble any of the observed relationships in Fig. 1. It is not surprising therefore that the partial molar volume plot of the tripositive lanthanoids resembles the reciprocal radius plot in Fig. 2. Consequently, electrostriction goes through a minimum in the middle of the series. This was accepted as evidence for a change in the primary hydration number between Nd and Tb, presumably from 9 to 8 (74). Recently, *Grenthe et al.* (75) have objected to this interpretation on the basis of evidence from partial molar heat capacity measurements of rare earth perchlorates. They conclude, in accord with earlier [17]O chemical shift data (76), that equilibrium (1) is not present for the hydrated RE ions in a perchlorate medium.

$$M(H_2O)_x^{+3} \rightleftharpoons M(H_2O)_{x-1}^{+3} + H_2O \qquad (1)$$

7

However, *Spedding* and co-workers (*74*) have just reported the results of X-ray diffraction studies of concentrated rare earth chloride and perchlorate solutions that demonstrate a nine coordinate primary hydration sphere for Pr^{+3}, while Tb^{+3}, Dy^{+3} and Er^{+3} are eight coordinate. The different ranges of concentrations examined in the various studies might explain these conflicting results.

According to the model of ionic hydration proposed by *Bockris* and *Saluja* (*77*), increased electrostriction implies larger numbers of waters of hydration in the hydration spheres of an ion. *Hepler* (*78*) has demonstrated that a linear relationship exists between ΔS^0 and ΔV^0 for ion association in aqueous solutions. Since most of the lanthanoid reactions in water are largely entropy stabilized (*70, 79*) — e.g., EDTA (*66*), carboxylic acids (*80*), acetylacetone (*67*) and diglyco-lates (*81*) — changes in electrostriction of water on complex formation are expected to be reflected in the free energies. This occurs in spite of a considerable compensation effect (*82*) due to contributions to ΔH^0 and ΔS^0 of the same sign (*79*). As explained in the next few paragraphs, the nature of this residual contribution of solvent effects to the equilibrium quotient may be understood more clearly if complex formation kinetics are examined.

A number of separate steps have been delineated in lanthanoid complex formation (*69, 83*).

$$\text{Ln (aq)} + \text{A(aq)} \underset{k_{21}}{\overset{k_{12}}{\rightleftarrows}} \text{LnW}_M\text{W}_A\text{A} \tag{2}$$

$$\text{LnW}_M\text{W}_A\text{A} \underset{k_{32}}{\overset{k_{23}}{\rightleftarrows}} \text{LnW}_M\text{A} + \text{W}_A \tag{3}$$

$$\text{LnW}_M\text{A} \underset{k_{43}}{\overset{k_{34}}{\rightleftarrows}} \text{LnA} + \text{W}_M \tag{4}$$

W_M and W_A denote the water molecules lost from the primary coordination sphere of the RE ion, Ln, and the ligand A, respectively. Reaction (2) involves a diffusion-controlled encounter, (3) desolvation of the anion and outer ion-pair formation, and (4) cation desolvation followed by entry of the ligand into the primary coordination sphere. The cation desolvation step is usually considered rate determining, although the opinion has recently been expressed (*83*) that a fourth step involving ring closure by a chelating ligand accompanied by further cation desolvation may well be the slowest.

Ultrasonic absorption techniques allow measurements of stepwise rate constants while most other methods yield only overall forward (k_f) or backward (k_b) rate constants.

$$\text{Ln(aq)} + \text{A(aq)} \underset{k_b}{\overset{k_f}{\rightleftarrows}} \text{LnA(aq)} \qquad K_1 = \frac{[\text{LnA}]}{[\text{Ln}][\text{A}]} = \frac{k_f}{k_b} \tag{5}$$

Normally, k_b is roughly constant for all RE ions with the same ligand. When reaction (4) is rate limiting it may be shown that $K_1 = c\,k_{34}$, with the constant $c = K_{12}K_{23}/k_{43}$, and K_{12} and K_{23} the equilibrium constants for steps (2) and (3).

8

It is expected that k_{34} should approximate the value of k_{ex}, the rate constant for water exchange. Consequently, the most stable complex in the RE series should also exhibit the most rapid solvent-bond rupture in reaction (4). The available water exchange rates (76, 83) also seem to exhibit an abrupt drop in magnitude between Gd^{+3} and Tb^{+3}. It is interesting that NMR shift studies of the state of hydration of Eu^{+3} in water-acetonitrile mixtures (84a) could only be analyzed successfully if two types of affinities were assigned to bound water molecules. Although the overall analysis was not critically dependent on the primary hydration number (8 or 9), the choice of binding constants was critical. Assuming the tricapped trigonal geometry for the fully hydrated species, the three equatorial sites had a much higher affinity for water ($K_e = 720$) than the six axial positions ($K_a = 120$). This result is somewhat surprising, as crystallographic data for the nonahydrates (Table 1) indicate that the equatorial bond-lengths are about 0.15 Å longer. Of course, free energy or equilibrium quotients need not reflect bond energies directly. Additional contributions from the entropy and enthalpy of lateral interactions in the primary coordination shell and of interaction with the secondary hydration sphere, as well as translational and vibrational entropy are also expected to determine the free energy of metal-ligand interactions (77). From a free energy point of view, the equatorial positions could be preferred in spite of somewhat longer bond-lengths, and at the same time be the most labile kinetically. In any case, the important conclusion is that all positions on the coordination polyhedron are not necessarily equivalent, and the structural data in Table 1 support this notion. This phenomenon provides a plausible interpretation for the characteristic and similar dependence on atomic number of k_{34} (83) and K_1 (Fig. 1) for the RE-acetate complexes. Shrinkage of a coordination polyhedron in a series of isomorphous compounds is known to accompany the reduction in the radius of the central ion, and may indeed be abrupt around the middle of the RE series (24). Perhaps this is what occurs for the water coordination polyhedron, and that in this process the positions of preferred binding and enhanced lability disappear (i.e., a reduction in k_{ex} and consequently in k_{34} and K_1). Such a process might involve a change in geometry or coordination number if the steric strain were severe enough.

A recent conformational study (see Section III, vi) of lanthanide complexes of indol-3-ylacetate in aqueous solution has shown that the carboxylate donor group acted as a bidentate ligand in the first half of the RE-series, and that the conformation of the complexes changed systematically along the members in the second half giving predominantly monodentate binding at Er^{+3} and Tm^{+3} (84b). For this ligand system, a change in reaction mechanism would therefore be expected around the middle of the lanthanide series. Bidentate binding has been rejected for RE-acetates (83) because of the goodness of fit of the kinetic data to the three-step model. Nevertheless, oxygen-17 shift data (76) provide some support that chelation may indeed be important. A change in the mode of binding need not necessarily affect the relationship between K_1 and k_{34}. If desolvation dominates the ring closure step, and if the second water molecule is more readily removed from the cation than the one in reaction (4), then the relationship $K_1 = ck_{34}$ still applies. For a simple planar charged donor group like the carboxylate function this is not an unreasonable expectation.

9

Ligand dependent chelate ring closure is believed to be rate determining for ligands such as oxalate, murexide and anthranilate (*83*). Equilibrium and kinetic data are plotted in Fig. 3 for oxalate and anthranilate. The behaviour of the

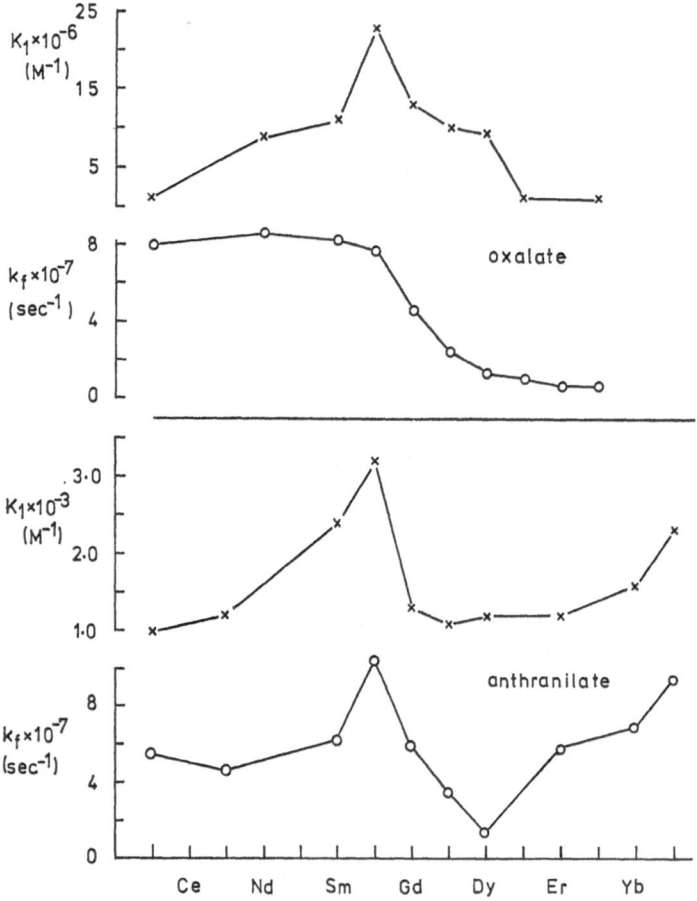

Fig. 3. K_1 and k_f values as a function of atomic number for oxalate (*85*) and anthranilate (*86*) complexes

murexide system parallels that of oxalate. In the case of anthranilate, the response of k_f and K_1 values across the series are similar, and k_b is approximately constant. As indicated above, invariance of k_b in a ligand system is a common observation in RE chemistry. Consequently, the factors that govern the magnitude of k_f (ion-pair formation and desolvation of anion, desolvation of cation, and ligand-dependent ring closure) are reflected in the magnitude of K_1. The equilibrium quotient behaviour pattern for oxalate and murexide resembles that depicted in Fig. 1 for acetate, although the sigmoidal relationship of k_f to atomic number does not resemble the k_{34} response observed for acetate. It appears that K_1 is still proportional to k_{34} or k_{ex} and that the chelation step dependence of k_f cancels out in the ration k_f/k_b.

One other phenomenon is know to influence lanthanide equilibria, and is summarized in Eq. (6).

$$LnA_j \ (H_2O)_x \ \rightleftharpoons \ LnA_j \ (H_2O)_y + (x-y)H_2O \qquad (6)$$

Spectrophotometric (66, 87–90), heat capacity (91, 92), polarographic (93), ^{18}O exchange (94), and NMR (95, 96) data provide convincing evidence that early in the RE series the EDTA complex exists predominantly as $Ln(EDTA) \ (H_2O)_x$ and as $Ln(EDTA) \ (H_2O)_y$ toward the end with $x-y=1$; for ions in the middle of the series, e.g., Eu^{+3}, both are present simultaneously. Equilibrium (6) is also believed to be important for the Ln (diglycolate)$\bar{_2}$ species (81, 97). We will briefly return to the EDTA system in a later section.

In conclusion, equilibria and kinetic studies of complex formation in ligand systems for which the binding site is not known may help in the elucidation of the nature of the binding groups. A detailed examination of available information for model systems, seems in order to establish the validity of this proposed diagnostic probe. Data for one enzyme system, staphylococcal nuclease (98), are encouraging.[3] The inhibition constants determined for the trivalent lanthanide ions in the calcium dependent DNA assay at pH 7 may be considered to reflect the reciprocal of the formation constant with nuclease in the presence of the substrate DNA. This value of K (complex formation) goes through a maximum in the middle of the series[4] and this suggests, in analogy with the acetate data in Fig. 1, that carboxylate donor groups play a dominant role in the complex formation. Crystallographic data of a nuclease-calcium-inhibitor complex (51, 52), confirm that there are three carboxylate groups in the primary coordination sphere of calcium, a peptide carbonyl, as well as a fourth carboxylate function at somewhat greater distance. The competive inhibition observed in this system implies that the tripositive lanthanides and calcium do occupy the same binding site.

iii) Electronic and Magnetic Properties

With few exceptions the trivalent "lanthanide ions have ground states with a single well-defined value of the total angular momentum, J, with the next lowest J state at energies many times kT (at ordinary temperatures equal to $\sim 200 \ cm^{-1}$) above and hence virtually unpopulated" (63). In the case of Sm^{+3} and Eu^{+3} the excited J states are appreciably populated at ordinary temperatures. For these ions the magnetic moments are not given by the well-known ground-state-only formula (7) (63, 99–102), but thermal population of the excited states as well as the second-order Zeeman effect must be considered in the calculation of the expected magnetic susceptibilities or moments.

$$\mu_{eff} = g[J(J+1)]^{1/2} \qquad (7)$$

---

[3] Equilibrium and enzyme activity data are compared in Section II for several other systems.
[4] The reciprocals of the inhibition constants determined were, 0.42 $(\mu M)^{-1}$ for La, 1.4(Pr), 1.7(Nd); 2.0(Gd); 1.7(Dy) and 0.46(Yb).

The observed effective magnetic moments when plotted as a function of atomic number resemble the curve obtained for the reciprocal ionic radius in Fig. 2. Values increase from zero at La^{+3} to ~ 3.3 BM at Nd^{+3} and then fall to ~ 1.5 BM at Sm^{+3}. Past Eu^{+3} (3.4 BM) the magnetic moment increases and reaches a maximum value at Dy^{+3} and Ho^{+3} (~ 10.5 BM), and then decreases again to zero at Lu^{+3} (63, 102).

Virtually all the absorption bands observed in the visible and near-ultraviolet spectra of trivalent lanthanoids exhibit a line-like character and are of low intensity (63, 103, 104, 105). Typically, molar absorptivities have values $\varepsilon \leq 6$ litres/mole-cm in aqueous solutions. The sharpness of the peaks indicate that the f-orbitals are well shielded from the surroundings and that the various states arising from the f^N configurations are split by external fields only to the extent of several hundred reciprocal centimeters. In addition, the intra-f^N transitions are formally parity forbidden by the rules of quantum mechanics. Various mechanisms have been proposed that could cause the breakdown of the selection rules, and these lead to electric-dipole (involving coupling with states of opposite parity), magnetic-dipole and electric-quadrupole transitions. Of special significance are the "hypersensitive" transitions which are believed to be pseudoelectric quadrupole in origin (105—107) and usually obey the selection rule $|\Delta J| = 2$. These are more sensitive to their environment than the normal f—f transitions, and should thus be useful in probing the immediate environment of the lanthanide ions such as in complexes. Another spectral feature of interest that has been observed and could be exploited in structural probe studies is ligand-to-metal charge transfer transitions (86, 108—110). These are important for those lanthanoids with the most stable divalent states (Eu, Sm, Yb, Tm) and tetravalent state (Ce). Studies with halides (LnX_3 and LnX_6^{-3}) also provide evidence for $4f \to 5d$ transitions (68, 110—112).

The fluorescence of rare-earth chelates has been the subject for extensive investigations in connection with its use in liquid lasers [e.g. (113—115)]. Almost all solution work has been done in organic solvents with chelates of the β-diketonate type. Complexes of Eu, Tb, Dy and Sm would appear to possess the most favourable fluorescence properties (56, 115). Emission line intensities are low unless energy transfer from a triplet level on the ligand to the f^N manifold of the rare-earth ion is possible. However, in the absence of such intersystem crossing from the ligand, chelation may still enhance some emission lines. For example, on complex formation with EDTA in an aqueous medium, the fluorescence of the $^5D_0 - {}^7F_2$ transition of Eu^{+3} increases approximately 100-fold (116). Improved intensities are possible for Tb^{+3} and Eu^{+3}, roughly seven- and twenty-fold respectively, by working in D_2O rather than H_2O (117). Hypersensitivity may again be employed to probe changes in the intermediate environment of the metal ion. It is noteworthy that the splitting observed in the $^5D_0 - {}^7F_0$ region in the absorption spectrum of EuEDTA (88, 89), also shows up clearly in the high-resolution fluorescence spectrum; and indeed was interpreted by Charles et al. (118) as signifying the presence of two isomeric forms (vide supra).

Perhaps the principle of the most versatile application of the rare-earth ions as fluorescent probes is demonstrated best by summarizing the work of Charles et al. (118, 119). At concentrations where the Tb^{+3} and the $Tb(EDTA)^-$ fluorescences were negligible, the mixed complex with EDTA and 5-sulphosalicylate (I),

Tb(EDTA)(SSA)$^{-4}$, showed intense fluorescence of high quantum efficiency when excited with near UV radiation. (The formation of the mixed species from the EDTA complex was measured to have a formation constant of 4.1 log units.) No such enhancement was observed for the corresponding Eu^{+3} system. However, in the mixed europium complex with EDTA and p-benzoyl-benzoate (II) near-ultraviolet radiation (360—400 nm) was absorbed by the secondary ligand and transferred to the Eu^{+3} for re-emission as the visible red fluorescence characteristic of this ion. In the SSA adduct of Eu(EDTA)$^-$, the very low fluorescence properties were attributed to a mismatch in the energy of the ligand triplet level and the europium resonance level. This strict energy compatibility limits the number of ligands that are able to transfer absorbed radiation to the central lanthanide (III) ion for fluorescence emission. This specificity may well be a blessing in disguise (vide infra).

I   II

iv) Dipolar and Contact Shift Properties

The most firmly established application of RE ions as structural probes is in NMR spectroscopy. Induced chemical shift perturbations usually depend on the type of nucleus examined and on the specific central ion in a paramagnetic complex. A distinction must be made between *contact* and *pseudocontact* shifts. The former may be viewed as a through-bond effect, while the latter arises because of a through-space interaction. The hyperfine contact intereaction between a nucleus in a ligand and the unpaired electron spin density on the central lanthanide is believed to occur because of spin polarization, rather than direct delocalization of the unpaired electrons (10). This conclusion is consistent with the accepted notion that 5f orbitals are too well shielded to participate in direct orbital overlap with ligand molecular orbitals. For the lanthanides, the fundamental relationship describing the contact shift (10, 120, 121) is given in Eq. (8).

$$\frac{\Delta v^{con}}{v_0} = \left(\frac{-A}{\hbar}\right) \frac{g_J\beta(g_J - 1) J(J + 1)}{3\,kT\gamma_I} \tag{8}$$

The symbols in this expression are defined as follows: Δv^{con}, the induced shift in Hz; v_0, the nuclear Larmor frequency in Hz; g_J, the Landé g-factor; $A/\hbar$, the scalar electron-nuclear hyperfine interaction constant in Hz; J, the total angular momentum quantum number; β, the Bohr Magneton; γ_I, the nuclear magneto-gyric radio; and k and T have their usual significance. Relative contact shifts for nuclei in the same molecule and of the same fractional spin occupancy are estimated to be (10): 1.0 (^{1}H), 9(^{13}C), 15(^{14}N), 18(^{31}P), 24(^{17}O), 36(^{19}F). The contact mechanism is thus of least importance for protons. *Reuben* and *Fiat* (10, 76, 122) have confirmed, albeit tentatively, that for the protons of RE bound water mocules the contact interaction was secondary importance. On the basis of this work and other "oversimplified" calculations (10), the greatest interaction is predicted for Eu^{+3}. This view is in direct opposition to the work of *Haas* and *Navon* (84a) which indicated that the pseudocontact model could adequately account for the observed water-proton shift patterns in water-acetonitrile mixtures. In most ligand systems the protons examined are not directly attached to the donor atom as in water. If a ligand is further specified to be saturated or unconjugated, we would expect the proton hyperfine coupling constant to be small. Indeed, this is the conclusion reached for substrate ^1H shift studies with the RE β-diketonates (9—13, 120), although significant contact interaction appears to occur for hydrogen nuclei ortho to the N-donor atom in pyridine related substrate systems. Large contact shifts have been assigned to all proton shifts of substituted pyridines induced by interaction with Pr^{+3} and Nd^{+3} salts in acetonitrile (18, 123). However, the data analysis for this system has recently been criticized (10).

Pseudo-contact interactions are the result of electron dipole-nuclear dipole interactions which are not averaged to zero by the rapid tumbling of a paramagnetic complex in solution if its magnetic susceptibilities are anisotropic. A generally valid expression for the fractional shift in a nuclear resonance frequency is given in Eq. (9), (120).

$$\frac{\Delta \nu^{\text{dip}}}{\nu_0} = - D \left[\frac{3 \cos^2 \Theta - 1}{r^3} \right] - D' \left[\frac{\sin^2 \Theta \cos 2 \Omega}{r^3} \right] \tag{9}$$

$$D = \frac{1}{3 N} \left[\chi_z - \frac{1}{2} (\chi_x + \chi_y) \right] ; \quad D' = \frac{1}{2 N} [\chi_x - \chi_y]$$

Under conditions of axial symmetry, $\chi_x = \chi_y = \chi_\perp$ and $\chi_z = \chi_\parallel$. Consequently, $D_{\text{ax}} = \chi_\parallel - \chi_\perp$ and $D'_{\text{ax}} = 0$. The symbols r, Θ and Ω are the spherical polar coordinates of the resonating nucleus in the coordinate system of the principal magnetic axes, the χ's are the principal molecular susceptibilities, and N is Avogadro's number. *Horrocks et al.* (120) have shown that RE magnetic anisotropies have both first order Zeeman (FOS) and second order Zeeman (SOZ) contributions. Theoretical calculations using values of phenomenological crystal field parameters for the axially symmetric compound $Yb(OH_2)_9(C_2H_5SO_4)_3$ indicate that the SOZ contribution to $\Delta \chi = \chi_\parallel - \chi_\perp$ dominates, although a non-negligible FOZ term was important for all three ions examined (Yb^{+3}, Pr^{+3}, Eu^{+3}). These findings do not agree with Bleaney's treatment (124) of the variation of magnetic anisotropies across the series. *Horrocks et al.* also compared the temperature dependencies of the calculated $\Delta \chi$ values with those of experimentally determined principal crystal susceptibilities for the corresponding $Ln(thd)_3(4\text{-picoline})_2$ complexes. These reciprocal temperature relationships were complex. However, the conclusion was reached that reasonable good T^{-1} plots may be expected for $\Delta \chi$ (and thus $\Delta \nu^{\text{dip}}/\nu_0$) "over the limited temperature range available to dipolar shift experiments, but that such plots will not in general pass through the origin". This conclusion is consistent with the bulk of the empirical data, and is in contrast to the T^{-2} dependence predicted by *Bleaney* (124, 125). The predominantly dipolar nature of proton resonance shifts in the β-diketonate systems has been convincingly established by comparing the signs and magnitudes of shifts with those expected from single crystal magnetic anisotropy measurements made on Ln(thd)$_3$. (4-picoline)$_2$ for ten RE ions (120, 126).

Solid state data [e.g. (29, 30, 40—42, 120)], as well as the single crystal suscept- ibility measurements alluded to above, suggest that the usual assumption of structural and/or magnetic axiality in solution is not a good one. However, the observed metal-ion independence in both aqueous (127—129) and organic solvents (9—13) of shift ratios for protons of the same substrate molecule and their agree- ment with values calculated from geometric parameters obtained from molecular models (9, 84a, 130) both overwhelmingly support the axial-symmetry assumption. Horrocks (131) has recently proposed a model which seems to account for this apparent or effective axiality. It is based on the assumption that 30 or more fluxional isomers interconvert rapidly on the NMR time scale. The stereochemical lability and abundance of geometric isomers for polyhedra of high co-ordination number ($\geq$7) is well established (15, 120). Even though the full form of Eq. (9) may apply for each individual isomer, calculations simulating isomeric intercon- versions indicate that Eq. (9) reduces the the simpler axial form ($D'=0$, $D=\chi_{||} - \chi_{\perp}$) when averaged for a large number of species ($\geq$30). Recently, Flora and Nieboer (132) have suggested that rapid intermolecular collisions or ligand exchange compete with such internal isomerization processes in reducing the effective anisotropy. This was necessary to explain the upfield shift away from the diamagnetic position with increasing temperature of the t-butyl methyl resonance of Eu(thd)$_3$ proper, and the dramatic shift to the diamagnetic position on dilution in a concentration range (0.005 — 0.04 M) where osmometric measurements in- dicated no decomposition of the shift reagent. Spectrophotometric measurements showed that in the same concentration range Beer's law was non-linear.

Under conditions of rapid exchange (vide infra) between uncomplexed ligand and metal-bound ligand, the observed shift ($\Delta\nu$) is related to the induced shift corresponding to the fully formed complex ($\Delta^*\nu$) through the fraction of complexed ligand p_L.

$$\Delta\nu = p_L \, \Delta^*\nu \tag{10}$$

$\Delta^*\nu$ values are usually obtained from plots of $\Delta\nu$ versus C_M/C_L, where C_M and C_L are the total metal and ligand concentrations, or from related functions. Elab- orate methods of analyzing organic shift reagent data have been summarized by Reuben (10), and should be applicable to studies in aqueous media. A novel and versatile approach to such equilibrium studies will be mentioned in a subsequent section.

v) Relaxation Enhancement Properties

This property is limited to Gd^{+3} and Eu^{+2} which have $^8S_{7/2}$ electronic ground states ($J=S$ as $L=0$). For the other RE ions the process is complicated by con- tributions due to anisotropic g-tensors (10, 133, 134). The relaxation rates for protons of a metal-bound ligand molecule are summarized in Eq. (11) and (12) (134—136).

$$\frac{1}{T_{1M}} = \frac{2}{15} \frac{\gamma_I^2 \, S(S+1) \, g^2 \, \beta^2}{\gamma^6} \left(\frac{3\tau_c}{1 + \omega_I^2 \, \tau_c^2} + \frac{7\tau_c}{1 + \omega_s^2 \, \tau_c^2} \right)$$
$$+ \frac{2}{3} S(S+1) \, (A/\hbar)^2 \left(\frac{\tau_e}{1 + \omega_s^2 \, \tau_e^2} \right) \tag{11}$$

$$\frac{1}{T_{2M}} = \frac{1}{15} \frac{\gamma_I^2 \, S(S+1) \, g^2 \, \beta^2}{\gamma^6} \left(4\tau_c + \frac{3\tau_c}{1 + \omega_I^2 \, \tau_c^2} + \frac{13\tau_c}{1 + \omega_s^2 \, \tau_c^2} \right)$$
$$+ \frac{1}{3} S(S+1) \, (A/\hbar)^2 \left(\tau_e + \frac{\tau_e}{1 + \omega_s^2 \, \tau_e^2} \right) \tag{12}$$

The quantities γ_I, β, r, S, g, A, h were defined previously, ω_s and ω_I are the electronic and nuclear Larmor precession frequencies, and τ_c and τ_e the correlation time for dipolar interaction and isotropic spin exchange respectively. T_{1M}^{-1} and T_{2M}^{-1} denote the longitudinal and transverse nuclear relaxation rates in a complex containing bound paramagnetic ions. The first term on the right-hand side of both Eq. (11) and (12) represents the proton-electron spin dipole-dipole inter-action, while the second term (that including A) the isotropic proton-electron spin exchange. The constants τ_c and τ_e characterize the rate at which the two forms of magnetic interactions between the ion and the proton are interrupted; the fastest mechanism is always the dominant one for isotropic motion. Thus,

$$1/\tau_c = 1/\tau_s + 1/\tau_M + 1/\tau_R \tag{13}$$

$$1/\tau_e = 1/\tau_s + 1/\tau_M \tag{14}$$

where, τ_M is the lifetime of a nucleus in the bound site which is determined by the rate of chemical exchange, τ_R is the rotational correlation time of the solvated paramagnetic complex, and τ_s is the electron spin relaxation time.[5] Temperature variation studies of observed T_1^{-1} and T_2^{-1} values [see Eq. (15)] in simple ligand systems of Gd[+3] (135, 138, 141), and perhaps of Eu[+2] (135), reveal that the scalar terms of Eq. (11) and (12) are insignificant. There is also general agreement that τ_R is the dominant correlation time in these systems (134, 135, 138, 142). Con-ditions of rapid exchange ($\tau_M \ll T_{1M}$, T_{2M}) would appear to prevail even when the ligand is macromolecular (7, 128, 129, 134, 135, 138). For a system in which rapid exchange between bound and unbound ligand has been established, the observed relaxation rates in the presence of the paramagnetic ion, $1/T_1$ and $1/T_2$, are related to the fraction of bound ligand p_L in the manner of Eq. (15).

$$T_x^{-1} = p_L(T_{xM}^{-1}) + (1 - p_L) \, (T_{x,0}^{-1}) \tag{15}$$

$T_{x,0}$ is the relaxation rate in the absence of the paramagnetic ion; $x = 1$ or 2.

When a paramagnetic ion binds to a macromolecule τ_R will increase and, if τ_R makes a contribution to τ_c initially, the value of τ_c (and thus $1/T_{xM}$) will also

---

[5] A distinction has been made between the longitudinal and transverse electron relaxation times (137—139). Koenig (140) and Dwek (134) have pointed out that in general this distinc-tion need not be made.

increase on binding. The relaxation enhancement factor ε^* is the normal empirical quantity that is used to measure this effect for protons of water.

$$\varepsilon^* = \frac{1/T_x^* - 1/T_{x,0}^*}{1/T_x - 1/T_{x,0}} \tag{16}$$

The asterisk denotes the presence of a macromolecule. Eq. (17) defines the enhancement factor ε_b^* for the hydrated metal-macromolecular complex. It may be calculated (18), from ε^* and X_M^*, the fraction of metal ion bound to the macromolecule which is readily obtained if C_L, C_M and the binding constant for the complex formation reaction are known.

$$(q^*/T_{xM}^*) = \varepsilon_b^* (q/T_{xM}) \tag{17}$$

$$\varepsilon^* = X_M^* \varepsilon_b^* + (1 - X_M^*) \tag{18}$$

Here, q represents the primary hydration number of the aquo-ion, and q^* that of the macromolecule-bound metal ion. Conversely, the binding constant and ε_b^* may be evaluated by curve-fitting procedures from plots of $1/\varepsilon^*$ versus C_L^{-1} (135).

In conclusion, the lanthanide ions are conveniently classified according to their intrinsic dipolar relaxation enhancement and dipolar chemical shifts properties (9, 10, 120, 126, 127):

a) The diamagnetic cations La^{+3} and Lu^{+3}, which are useful in evaluating non-paramagnetic contributions to the parameters observed.

b) Cations from the first half of the RE series, Pr^{+3}, Nd^{+3} and Eu^{+3}, which induce little line broadening because of a combination of short electron-spin relaxation times and moderate magnetic anisotropies. These anisotropies are large enough so that chemical shift perturbations are readily observed.

c) Cations from the second half of the series, Tb^{+3}, Dy^{+3}, Ho^{+3} and to a lesser extent Er^{+3}, Tm^{+3}, and perhaps Yb^{+3}, have somewhat longer electron-spin relaxation times and relatively large magnetic anisotropies. Although these ions cause appreciable line broadening, they do provide considerably larger chemical shift perturbations.

d) The S electron ground state ions Gd^{+3} and Eu^{+2} that possess long electron spin relaxation times; long compared to those of the other RE ions and to their own rotational correlation time, τ_R. They are responsible for pronounced isotropic line broadening.

e) Pr^{+3}, Nd^{+3}, Tb^{+3}, Dy^{+3} and Ho^{+3} cause shifts in the opposite direction from those induced by Eu^{+3}, Er^{+3}, Tm^{+3} and Yb^{+3}.

vi) Miscellaneous Physicochemical Properties

There are known properties of specific RE ions that have not been extensively employed in probe studies. These are now summarized.

EPR. The long electron relaxation times of Gd^{+3} and Eu^{+2} allow the observation of their electron paramagnetic resonance (EPR) spectra in solution (135, 139). Reuben (139) found that line-widths and the characteristic correlation times were sensitive to complex formation with dimethylarsinate (cacodylate) and the protein bovine serum albumin. The similarity of the results obtained to those of Mn(II) suggests that EPR may be used in the manner documented for this ion (143—145 a). An innovative approach to determine the relative positions of photosynthetic election carriers in the chromatophore membrane in *Chromatium* involved the dipolar perturbation by externally bound Gd^{+3} of the EPR spectra of the electron carriers (145 b).

Mössbauer. Europium-151 (47.8% natural abundance) is amenable to Mössbauer spectroscopy. Isomer shifts for Eu(II) are considerably larger and more sensitive to the ion's environment than those for Eu(III), which are small and insensitive (6, 146, 147 a, b). The author's extensive experience in handling Eu^{+2} solutions [cf. (6, 135)] confirms that extremely anaerobic experimental conditions are imperative if the divalent state is to be retained. However, many solid Eu(II) compounds are stable if kept dry [e.g., $EuSO_4$, $EuCO_3$, $Eu(HCO_2)_2$], and thus solid derivatives of such compounds as calcium proteins should be feasible for Mössbauer measurements. Mössbauer studies at 4.2 °K of Eu^{+3}-transferrin (147 c) revealed a singlet without any fine structure. This manifestation of a total lack of sensitivity to environmental parameters was related to the singular population of the diamagnetic electric ground state $^7F_0$ at low temperatures.

Polarography. The stable, but strongly oxidizing, tetravalent state of cerium and the relatively stable divalent state of europium may be studied by electrochemical methods (148—150). For example, polarographic studies of the Eu(EDTA) system not only indicated the presence of two different isomeric forms [cf. (6)], but also allowed the evaluation of the various equilibrium constants (93). The different half-wave potentials observed for the two isomers showed that this parameter is sensitive to the immediate environment of the metal. It may be significant that the standard potential of the couple Eu^{+3}/Eu^{+2}, $E° = -0.43$ V, becomes positive as the water content is lowered in water-acetonitrile mixtures (151).

vii) Analytical Chemistry

Many instrumental methods are available for the quantitative analysis of the rare earth elements. These range from spectrophotometric analysis (103, 149), to atomic absorption and emission (instrument manufacturer's manual gives detail), X-ray fluorescence (149, 152) and photoluminescence (115) techniques. In passing, it is of interest to note that the SSA-sensitised Tb(EDTA) fluorescence (vide supra) allows the determination of micro-quantities of Tb^{+3}. A sensitivity down to $2 \times 10^{-8} M$ is claimed (153). Less general instrumental methods are available for individual lanthanides. Thus europium is readily determined by polarography (148, 149), or coulometric oxidation of Eu^{+2} preceded by reduction of Eu^{+3} at a mercury cathode of controlled potential (154, 155). Since most tripositive lanthanides are purchased at a purity of 99.9% as chlorides, nitrates or oxides, less elaborate methods of analysis than those mentioned are normally needed. Those commonly employed are described below.

Lanthanide solutions must be kept slightly acidic to prevent hydrolysis, since the RE hydroxides are only very slightly soluble (~ 1 μM), (149). This problem is especially acute in concentrated solutions (*e.g.*, 10^{-1} M). Complexiometric titration with EDTA is by far the most convenient standardization procedure. Xylenol orange is a suitable indicator (*149*). In the authors' laboratory, a back-titration technique with a standard copper solution and PAN as indicator is frequently employed (*156*). Eu_2O_3 (99.9% pure) has been shown to be a good primary standard after firing at 900 °C for 1—2 hours (*96*). Presumably this procedure applies for the other oxides as well. A gravimetric method with some advantages is the oxalate precipitation. Since traces of transition paramagnetic ions are removed in this process, it may also be used preparatively. The oxalate salt is readily converted to the oxide (*149, 157*), or may be titrated with potassium permanganate after dissolution in dilute sulphuric acid. Removal of traces of calcium from lanthanide samples may be affected by preferential precipitation of the RE ions as hydroxides (*149*). Residual calcium could be a source of error in assays of calcium enzymes (*158*). Finally, there is some merit in working with the perchlorate salts of the lanthanides, as there is evidence that chloride, nitrate and sulphate (has relatively low solubility) do interact with the ions in solution, likely by an outer-sphere mechanism (*76, 83, 159*). This interaction with the anion will be most significant when working in rather concentrated solutions. The perchlorates are readily prepared from the respective oxides by dissolving them in perchloric acid (*e.g.* 2 M), followed by repeated evaporation to dryness under an IR lamp to remove the last traces of acid (*160, 161*). Stock solutions thus prepared, and adjusted to a suitable pH (*e.g.*, pH 3), are readily standardized by passing aliquots through a strong acid cation exchange resin (Dowex-50, or Amberlite IR-120 H) and titrating the eluant with standard base.

II. Biological Activity

The lanthanide ions have no known inherent biological function, and only trace amounts are found in whole body analyses (*162*). Since the RE ions are classified as class "a" acids, and have sizes (0.85 — 1.06 Å) comparable to Ca^{+2} (1.06 Å)[6], it is not surprising that biological activity is observed in a number of calcium enzymes. They have been demonstrated to be good substitutes for calcium in assays of *Bacillus subtilis* α-amylase (*158, 163, 164*), of trypsinogen to trypsin conversion (*165, 166a*) and of the aequorin luminescent reaction (*166b*); and for Mg^{+2} in isoleucyl-tRNA synthetase (*167*) and adenylylated glutamine synthetase (*168*). The enzyme activities of the α-amylase (relative to Ca^{+2} as 100%) and the tRNA synthetase are shown in Fig. 4 for most of the RE ions. *Kayne* and *Cohn* (*167*) have demonstrated that the role of the lanthanides, and presumably divalent ions other than Mg^{+2}, was to stabilize the substrate tRNA in the overall reaction. Mg^{+2} catalyzes the formation of the ileu-AMP-enzyme complex as well, in contrast to the lanthanide ions. Since it is expected that Mg^{+2} retains its waters of hydration when bound to RNA or DNA (*6*), some correlation of enzyme activity and "outer"-complex binding might be expected. It is indeed encouraging that

---

[6] This is the value of Goldschmidt; Pauling's value is 0.99 Å [Ref. (*63*), p. 52].

the enzyme activity pattern across the series correlates well with that of lanthanide-nitrate interaction (83, 159), which is known to be of both the "inner" and "outer" type (83, 169, 170). Thus, interaction with partial loss of waters of hydration would appear to be a good model for the cation stabilization of polynucleotides, at least under the experimental conditions of the isoleucyl-tRNA synthetase assay.

Similarly, the agreement between B subtilis α-amylase activity and porcine trypsin lanthanide complex formation is striking. Assuming that α-amylase activation is proportional to the fraction of bound metal (i.e., the binding constant), one is tempted to conclude that the binding in these two enzyme systems is very similar. Speculations based on the three dimensional structure of bovine trypsin

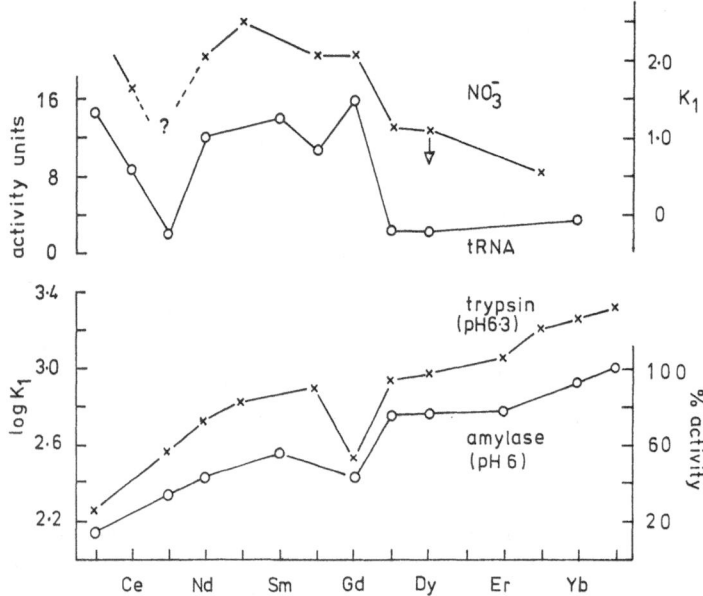

Fig. 4. Reactivation of *Bacillus subtilis* apo-α-amylase (158), isoleucyl-tRNA synthetase activity (167), and log K_1 values for nitrate (83, 159) and porcine trypsin (194a) as a function of atomic number. Values of log K_1 (MNO$_3^{+2}$) in the case of Nd, Gd, and Dy were corrected to correspond to an ionic strength of 1.0 M; the Dy value is an upper limit (83)

at 2.7 Å resolution suggests that Ca^{+2} binds at a site where several carboxylate side chains come together (171)[7]. Ca^{+2} is also known to compete effectively for one of the two Gd^{+3} binding sites of B subtilis α-amylase (163), presumably the one essential for the functional integrity of the enzyme. The trypsin and α-amylase curves in Fig. 4 reflect the reciprocal radius and $- \Delta \bar{V}^0$ plots in Fig. 2. The good agree-

---

[7] The primary sequence of porcine trypsin suggests that this particular chelate structure of bovine trypsin is absent. A strict comparison of the two enzymes may therefore be illusory. *Hermodson, M. A., Ericsson, L. H., Neurath, H., Walsh, K. A.*: Biochemistry 12, 3146 (1973).

ment (vide supra) of the staphylococcal data with that of carboxylic acid lanthanide complexes [c.f., log K_1 and log β_2 ($\beta_2 = K_1 K_2$) data for acetate (64), propionate (80) and isobutyrates (80)] suggests that the metal in these two enzyme systems is not at a simple carboxylate site, but perhaps at one modified by the presence of a metal alcohol linkage [cf., stability data for glycolate and lactate (172)].

Another significant correlation can be established between enzyme activity and kinetic and equilibrium data from model systems. A plot for trypsinogen of "half time" of conversion (time taken to reach one half of the maximum conversion rate at lanthanide ion concentrations for maximal zymogen conversion), or that of % activity (165, 166a, 173), versus atomic number almost reproduces the k_t (oxalate) curve in Fig. 3 and of the free energy of formation of [Ln(hist H)]$^{+3}$, (160). Darnall et al. (173) indeed postulate that the Ca^{+2} chelates with two carboxyl groups and reduces the negative charge in the area of the positively charged lysine of trypsinogen so as to maximize its interaction with trypsin.

The now resolved controversy about RE activation of B. subtilis α-amylase (158, 163, 164) emphasizes the need in assay studies to be prudent in the selection of buffers and RE ion concentration ranges. The biological activity of the lanthanide ions is critically dependent on the "free" available metal ion concentration in the assay mixture. Too high levels of metal are known to be inhibiting.

Competitive inhibition has been established for several calcium enzymes: staphylococcal nuclease (98) and concanavalin A (174, 175). Because of their extra charge and affinity for higher co-ordination numbers, the RE ions could distort the metal ion co-cordination polyhedron enough to destroy the catalytic ability of an enzyme. This could explain why Ca^{+2} enhances the binding of the transition metal in an adjacent site of concanavalin A, while the lanthanides do not (175); and, why one RE ion occupies two adjacent Ca^{+2} sites in thermolysin (53, 54). Similarly, the different spatial requirements of Mg^{+2} and its much slower (≥ 2 log units) rate of inner sphere substitution, compared to the lanthanoid ions [Ref. (63), p. 657], adequately account for the loss of biological activity when Mg^{+2} is replaced in such proteins as pyruvate kinase (176) and yeast inorganic pyrophosphatase (177). The lack of activity when the lanthanides are substituted for first row transition divalent metal ions, as in concanavalin A (175) and alkaline phosphatase (194e), is not surprising as the ligand type and geometric requirements of "borderline" (more class b character) binding sites normally do not match the coordination requirements of the class a lanthanide ions.

A number of functions and applications of the RE ions in physiological processes are summarized in Table 2. The list is not exhaustive, but is intended to indicate the manner and scope of lanthanide substitution for calcium in some very fundamental processes of life. The most obvious deduction from the data in Table 2 is that the differences between the responses to Ln^{+3} ions and Ca^{+2} are as important as the similarities. In most situations, the Ln^{+3} ions do not penetrate cell membranes but do inhibit Ca^{+2} influx and efflux. This difference in response has been employed in the study of various activators of molluscan catch muscle (182). It was possible to delineate that in activation by high concentrations of K$^+$, the major source of activating Ca^{+2} originated from outside the cell. In contrast, in activation by acetylcholine and neural excitation a significant portion of the Ca^{+2}

E. Nieboer

Table 2. Trivalent lanthanide ions in biological systems

System	Action or application	Ref.
Lobster axon	La^{+3} binds to the Ca^{+2} binding sites on the axon membrane and simulates membrane conductance and other calcium-like actions	(178)
Squid axon	La^{+3} reduces both the influx and efflux of Ca^{+2}	(179)
Mammalian smooth muscle	La^{+3} reduces both the influx and efflux of Ca^{+2}	(180)
Smooth muscle (guinea-pig ileum, rat uterus, rabbit aorta), striated muscle (frog sartorius and rectus abdominis muscles), and molluscan catch muscle	Replacement of Ca^{+2} with La^{+3} at sites accessible to the extracellular bathing solution provided information about Ca^{+2} movement and the mechanisms by which pharmacological agents act to initiate contractile responses in these systems.	(181) (182)
Frog twitch skeletal muscle	In hypertonic solutions La^{+3} enters cell. Such intracellular La^{+3} blocks movement of Ca^{+2} across membrane of terminal cisternae and is responsible for electromechanical uncoupling of tension and action potential.	(183)
Mammalian cardiac muscle	La^{+3} inhibited slow inward Ca^{+2} current with a concomitant decrease in contraction.	(184)
Sarcoplasmic reticulum of rabbit skeletal muscle	La^{+3} (Gd^{+3}, Yb^{+3}) were weakly competitive with Ca^{+2} for specific, high affinity, binding sites believed to be important in Ca^{+2} translocation.	(185)
Tumors	La^{+3} increased the mitochondrial uptake of calcium, in contrast to nonmalignant cells.	(186)
Rat liver mitochondria	La^{+3} binds to and inhibits the specific carrier system transporting Ca^{+2} inward, and is not transported itself.	(187)
Dog cardiac microsomes	La^{+3} does not affect Ca^{+2} accumulation, exchange or the Ca^{+2} stimulated ATPase	(188)
Bone proteins	La^{+3} competes with Ca^{+2} for the glutamic acid and aspartic acid residues	(189)
Corn roots	La^{+3} binds to cell walls and along plasma membranes up to the Casparian strip, and like Ca^{+2}, inhibits K^+ absorption; similar binding sites for La^{+3} and Ca^{+2} are inferred.	(190)
Secretory systems	La^{+3} alters Ca^{+2}-dependent secretory action. For example, La^{+3} stimulates the spontaneous release of histamine from mast cells, but is a potent inhibitor of the calcium-dependent component of antigen-stimulated histamine release.	(181) and references cited therein.

was released from internal sites on the sarcoplasmic reticulum. Similarly, studies with mammalian cardiac preparations (*184*) revealed that the slow inward current was predominantly carried by calcium because La^{+3} reduced the uptake of $^{45}Ca^{+2}$ and caused a concomitant decrease in the measured contractions. The inward current carried by sodium was further shown to be distinct from that carried by calcium. Consequently, in the above investigations the lanthanides were useful because they affected some but not all of a tissue's Ca^{+2} sites or stores.

Another interesting aspect of the studies summarized in Table 2 is that the specificity for Ca^{+2} in the mitochondrial ion transport system (*187*) differs significantly from that of the muscle sarcoplasmic reticulum (*185*) and cardiac microsomes (*188*). In the first case, La^{+3} was inhibitory while in the last two instances it was not at comparable concentrations. Eventhough the binding to specific high affinity sites was similar and dependent on energy-linked respiration for rat mitochondria (*187*), Ca^{+2} does and La^{+3} does not activate respiration. In contrast, the low affinity binding of Ca^{+2} was coupled to energetic processes while that of the Ln^{+3} was not. These sites were presumed to be different. It is tempting to speculate whether such divergent responses to the trivalent RE ions and Ca^{+2} have their origin in the small but real differences in physical and chemical properties (*e.g.*, charge, acidity in aqueous systems, co-ordination number, solubility and free energy of complex formation). If so, comparative complex formation and, where applicable, activity and biological response studies involving all the RE ions should assist in identifying the donor groups at the binding sites. This would be an enviable step forward in unravelling the complexities of calcium dependent life processes. As explained in Part III, some of the physical probe properties of the lanthanides are applicable to these heterogeneous biological systems.

III. Rare Earth Ions as Structural Probes

i) Heavy Atom Isomorphous Replacement Studies

The lanthanides are heavy metals and could be expected to be useful in electron microscopy and in X-ray studies. Since the Ln^{+3} ions do not penetrate membranes under normal conditions, they have been employed in the definition of extracellular channels in animals and the cell wall continuum in plants [(*190*), and references cited therein]. In corn roots, La^{+3} movement could be followed by electron microscopy along the cell wall continuum and the outside of the plasma membrane up to the barrier to solute diffusion provided by the Casparian strip (*190*). K^+ absorption studies showed that in this plant the binding sites for Ca^{+2} and La^{+3} were similar since both ions inhibited the inward movement of K^+. Electron microscopy employed in this manner would appear to be a good probe of cell membranes in both plant and animal organisms.

The direct determination of the phases of X-ray reflections by the method of multiple isomorphous replacement is the most powerful method of solving a protein crystal structure. Since some of the lanthanides possess large anomalous scattering properties, their use in phase determination is very favourable. Incorporation of these ions should occur readily even into proteins that are not metalloproteins or metal-protein complexes in their native form, because class "a" donor groups such as carboxylate anions are ubiquitous. When Gd^{+3} was employed as the heavy atom in three-dimensional studies of egg white lysozyme (7, 191), it bound at a single site between the two carboxylate residues of GLU 35 and ASP 52 which define the active site of this enzyme. The spectroscopic studies based on this Gd^{+3} reporter site will be reviewed presently. In the structure determination of the Fe-S complex in a bacterial ferredoxin (192), the usual heavy atom reagents (Hg or Pt derivatives are common) failed to yield a suitable derivative. However, UO_2^{+2}, Sm^{+3} and Pr^{+3} salts proved successfull.

A more specific and systematic investigation of the rare earths as isomorphous calcium replacements for protein crystallography is the work of Colman et al. (54) on thermolysin. Calcium is needed to stabilize the thermolysin structure but is not directly involved in its catalytic acitivity. X-ray analysis (53, 54) has identified four calcium sites, two single ones and a set separated by 3.8 Å. Electron density difference maps at 2.3 Å resolution between Ln^{+3} thermolysin and the native calcium form showed that the RE replaced Ca^{+2} at three of the four sites. One lanthanide ion occupied the two adjacent calcium sites. The replacement is accompanied by a minimal perturbation of the protein structure and the lanthanoid derivatives for this enzyme were superior to those incorporating Sr^{+2} and Ba^{+2}. However, the study showed quite clearly that in all three substitutions the lanthanide site differed from that of calcium by about 0.5 Å, and strongly suggested that the coordination was modified to satisfy the higher coordination number requirement of the lanthanides. Nevertheless, these perturbations were not severe enough to cause a serious loss in the thermal stability and proteolytic activity of this enzyme.

ii) Fluorescence Probe Studies

The fluorescence of Tb^{+3} when bound to certain proteins is enhanced by as much a factor of 10^4 or 10^5. Such enhancement has been observed for transferrin (193), concanavalin A (175), porcine trypsin (194a), bovine trypsin (142), conalbumin (194b) and, rabbit muscle troponin (194c, 194d). Enhanced terbium luminescence has also been reported for two other proteins, paravalbumin (194d) and alkaline phosphatase (194e), although the exact magnitudes of the increases were not reported. Enhancement factors of several hundred(s) have also been observed when terbium was bound to the antibody fragment of Myeloma Protein MOPC 315 (194f) and for complexes of Eu^{+3} and Tb^{+3} with Escherichia coli tRNA (194g). Luminescence studies of nucleotide-terbium complexes (194h) and nucleotide-europium complexes (194i) in aqueous solution are also known. The analogy of many of these protein systems with the SSA-sensitized Tb(EDTA) fluorescence discussed in Section I-(iii) and I-(vii) is striking. In all cases the excitation wavelength is in the UV region (300 ± 20 nm) and the fluorescence is greatly enhanced around 545 nm and to a lesser extent at 490, 590 and 625 nm. In common is also the lack of sensitization of the corresponding Eu^{+3} complex. The assignment of the absorbing donor group to a tyrosyl in the primary or secondary co-ordination sphere of the metal for four of these biological systems is in agreement with the known co-ordination of the phenolic SSA ligand to Tb^{+3} in its EDTA complex. A tyrosine carbonyl oxygen has indeed been assigned as one of the ligands at the calcium site of concanavalin A (50). In the case of transferrin (and conalbumin) the ultraviolet absorption difference spectrum of Tb^{+3}-saturated transferrin *versus* transferrin exhibits the characteristic shape of a difference spectrum due to tyrosinate residues of a protein (λ_{max} at 245 and 295 nm). The aromatic chromophore transferring its excitation energy to Tb^{+3} was identified as a tyrosine for troponin (*194c*) by comparing quenched (Tb^{+3} present) and unquenched intensities of the aromatic emission envelope. For porcine trypsin the sentitizing group was identified with neighbouring tryptophan residues assumed to be at a very short distance from the Ca^{+2} site. However, an excitation wavelength of 295 nm leads to the largest enhancement and is usually characteristic of an absorbing tyrosine group. The presence of a protonated tyrosinate group in the primary or secondary coordination sphere of Tb^{+3} should therefore not be discounted. [The formation of the trypsin-Tb^{+3} complex as a function of pH would seem to preclude the tyrosinate anion as ligand (*194a*).] The porcine trypsin stability sequence across the RE series (Fig. 4) discussed in Section II supports such an interpretation. This characteristic excitation wavelength at 295 nm has also been used to identify the involvement of a tyrosine residue in the binding of terbium to suspended erythrocyte ghosts from human blood and rat thymocytes (*194j*). An enhancement factor of three log units was observed. This application of terbium as a luminescent probe for Ca^{+2} in biomembranes is exciting and opens up many new avenues of study since Ca^{+2} is involved in numerous biomembrane phenomena such as intercellular communication, cellular adhesion, neurotransmitter release, and potassium transport.

Only in one instance has a chromophore other than the tyrosine side chain been positively identified. Parvalbumin is a protein that contains no tyrosine or tryptophan residues, and yet it exhibits intense terbium fluorescence (*194d*). The excitation wavelength was diagnostically low at 259 nm, corresponding to the spectral region where only phenylalanine absorption bands appear. Phe-57 and Tb^{+3} are known from X-ray studies (49, 194d) to be near one of the two calcium binding sites of parvalbumin. It is interesting that this phenylalanine residue matches with Tyr-108 in troponin C from rabbit muscle, which has been shown to be a protein homologous with calcium-binding parvalbumin. As indicated earlier, troponin C fluoresces in the presence of terbium with a λ_{ex} value of 280 nm, characteristic of the participation of a tyrosyl residue in the energy transfer process.

The work of *Kayne* and *Cohen* (*194g*) has shown the potential of both terbium and europium fluorescence enhancement studies for probing the structure of polynucleotides. Detailed studies with several tRNA molecules from *E. coli* and yeast have demonstrated that four to six lanthanide ions were bound firmly to the nucleic acid and that the uncommon 4-thiouridine unit was responsible for the increase in lanthanide luminescence [$\lambda_{ex} = 345$ nm; $\lambda_{em} = 545$ nm (Tb^{+3}), 585 nm (Eu^{+3})]. The absence of excitation bands in the region of common base residue absorption (*194k*), 250–300 nm, was assigned to strong absorption by tRNA, since Eu^{+3} (*194i*) and Tb^{+3} (*194h*) are known to be sensitized in this wavelength region by mononucleotides. In the latter study with simple nucleotides, both triplet and singlet state energy transfer from the aromatic base (donor) to the europium ion for subsequent emission were delineated (194i). The mechanism of energy transfer was believed to be of the collisional type, although the assumption used by the authors that no complex formation occured at pH 5 between Eu^{+3} and the mononucleotides has subsequently been disproven (127, 194h).

Tb^{+3} fluorescence is readily adapted to the evaluation of binding constants (175, 194a). Scatchard plots (195) are useful in the evaluation of the number and affinity of multiple sites. Competition studies involving the other RE ions and Ca^{+2} have been shown to be feasible, and were used to evaluate the porcine trypsin data in Fig. 4 (194a). In Section I-(iii) the enhancement of lanthanide fluorescence achieved when working in D_2O was mentioned. Luk (193) used this phenomenon to estimate the state of hydration of Tb^{+3} in the transferrin complex.

Luminescence intensities of various trivalent RE ions are sometimes enhanced or quenched by the coexistence of other kinds of trivalent lanthanoids or transition metal ions. Equations describing such transfer depend on the mechanism that operates in a system. The interaction between two RE ions is thought to be a dipole-quadrupole resonance transfer (193, 196), while a dipole-dipole mechanism (197) was assumed for RE-transition metal ion pairs (193). No quenching was observed with either combination (Tb^{+3}/RE, Tb^{+3}/Fe^{+3}, Tb^{+3}/Cu^{+2}) for the two binding sites of transferrin (193). In contrast to the Tb^{+3}/RE pairs, the combination Tb^{+3}/Fe^{+3} showed a measurable decrease in the fluorescence lifetime relative to the Tb^{+3}/Tb^{+3} complex. This reduction when combined with intensity measurements of overlapping acceptor (Fe^{+3}) absorption peaks and the donor (Tb^{+3}) fluorescence peak allowed the separation between the two metal centers to be evaluated (≥ 43 Å).

Circular polarization of terbium fluorescence has been detected in several cases. The relative intensity of the polarized light is compared to the total fluorescence intensity and the resulting ratio, the anisotropic factor, reflects the asymmetry of the chromophore in its excited state. The optical activity should therefore be very sensitive to the nature of the environment of the terbium in a complex. Consequently, the similar anisotropy factors and emission spectra observed for transferrin and conalbumin (194b) could be interpreted to signify similar metal-binding sites in these two proteins. Asymmetric environments for terbium may thus also be inferred from the distinct polarization of terbium emission detected for parvalbumin (194d), troponin (194c) and concanavalin A (194c).

iii) Difference Absorption Studies

It has already been mentioned that a few of the transitions out of the whole absorption spectra show a marked enhancement of intensity when water in the co-ordination sphere of the RE ions is replaced by a ligand. Katzin (107) concluded from studies with Eu^{+3} and sugar acids and amino acids as ligands that the greatest intensification occurred for the entry of the carboxyl group into the co-ordination shell. He observed also that the quantitative hypersensitive influence varied considerably with certain details of the structure of the anion. This concurs with the general observations of Birnbaum et al. (198, 199) based on work with amino acids and carboxylic acids at large ligand-to-metal concentrations (9:1).

Considerable improvement in sensitivity was achieved in this type of spectro-photometric studies by *Birnbaum et al. (166, 198—200)* who introduced the use of absorption difference spectra. They examined the changes in intensity and shape of the hypersensitive Nd^{+3} transition $^4I_{9/2} \to {}^4G_{5/2}$, $^2G_{7/2}$ employed by *Karraker (201)* to monitor the effect of adding chloride or perchlorate ions to aqueous solutions. The absorption occurs in the 560—600 nm portion of the visible spectrum. For the nitrogen-oxygen ligands examined there was no drastic change in the shape of the absorption difference peak, not even for EDTA. For 1:1 monodentate complex formation in mildly acidic solutions containing equal amounts of ligand and Nd^{+3}, it was found that the intensity (area under curve) of the difference peak correlated with the pK_a value of the carboxylate function or alternatively, the value of K_1 for the Nd-ligand interaction. (A similar relationship has recently been reported (194l, 194m) for the oscillator strength (is proportional to intensity) of the same f-f transition of Nd^{+3} and the basicity of eight ligands.) Estimates of relative intensities and ligand pK_a values, as well as formation constants (K_1) for monodentate interaction, are compiled in Table 3 for a number of carboxylic acids and amino acids. Intensity data for pH 7, where some of the ligands could potentially be bidentate or even tridentate, are also included (but see footnotes [f]) and [g]) to Table 3).

It is concluded from the data in Table 3 that metal-nitrogen interaction does contribute to the observed intensity even though the shapes of the difference

Table 3. Relative intensities for neodymium difference spectra
$\left(\begin{array}{l} ^4I_{9/2} \to {}^4G_{5/2},\ {}^2G_{7/2}\ \text{transitions}) \\ 570\text{—}600\ \text{nm} \end{array}\right)$

Ligand	pK_{a1}[a])	K_1[b])	Estimated intensities[c]) for monodentate complexes (cm²)	Estimated intensities[c,d]) for complex formation at pH 7 (cm²)	Likely[e,f]) denticity at pH 7
Histidine	1.82	1.8	4	50 (20)[g])	mono, bi, tri
Anthranilate	2.05	—	~5	80	bi
Alanine	2.35	6.5	9	16	mono, bi
Benzoate	4.2	—	45	45	mono
Acetate	4.75	83	50	50	mono

a) From (200).
b) From (200) and (202).
c) Estimated from Fig. 4 and 5 in Ref. (200); correspond to solutions 0.05 M in metal ions and ligand.
d) Hydroxide species were evident at pH >7.2 in some cases.
e) See Ref. (86), (160) and (203) for detailed assignments.
f) Hydrolysis has been shown (194n) to begin for lanthanide complexes with simple amino acids like alanine at about pH 6.5 and for histidine at about pH 7 by potentiometry and circular dichroism spectroscopy. Consequently, the deduction from increased intensities around pH 7 that the α-amino group binds to the metal is likely in error.
g) Value refers to pH 6.5 and bidentate denticity.

spectra are not affected. Difference spectra for trypsin (166a) and bovine serum albumin (198) had shapes identical to those recorded for the simple ligands. Consequently, the presence of nitrogen donors at the calcium sites in these proteins may not be excluded in an *a priori* fashion, although there is not much precedent for their presence especially not at pH 5.6 at which the spectra were recorded [cf., data for amino acids (200)]. No data is available for metal-phenol, nor unequivocal data for metal-alcohol, interaction [but see (107) and (199)]. It is significant that the intensity for the trypsinogen difference spectrum compared to that of trypsin is considerable larger for equimolar solutions (166a). This concurs with the knowledge of a second calcium binding site for trypsinogen. Studies with trypsin further revealed that the difference spectrum was sensitive to the *in situ* replacement of Nd^{+3} by Ca^{+2}. Finally, *Wedler* and *D'Aurora* (168) showed that binding of small molecules (substrates and modifiers) to glutamine synthetase of *E. coli* was reflected in the Nd-enzyme neodymium difference spectrum.

Circular dichroism (CD) spectra of the lanthanide ions would appear to be more sensitive to environmental ligand dependant parameters than ordinary absorption spectra (107, 194n). However, the weakness of the CD intensities may limit its application to low molecular weight ligand systems. In contrast, relatively strong magnetic circular dichroism (MCD) spectra are known for neodymium complexes in solution (194o). Sensitivities are such that complex formation between Nd^{+3} and parvalbumin could be detected (194d). Nevertheless, much more effort, both theoretical and experimental, is required before MCD (alternatively referred to as the Faraday effect) finds application as a useful structural probe.

iv) Relaxation Enhancement Probe Studies

A typical titration curve based on water proton relaxation enhancement spin-echo measurements (204) with Gd^{+3} is given in Fig. 5. ε_b^* values are either obtained by extrapolating such a curve to infinite protein concentration, or by an iterative procedure since ε^* is related to the equilibrium parameters through Eq. (18). The latter approach adjusts values of ε_b^* and the dissociation constant (K_d) for the complex until the best fit to the double reciprocal plot of Fig. 5 is obtained. The extrapolation procedure is often used in combination with the Scatchard plot (195) since the fraction of metal bound, X_m^*, can be calculated from ε^* and the ε_b^* value [cf. (18)]. Such plots readily reveal the number of metal-binding sites and their affinities. This approach has been employed in studying the binding of Gd^{+3} to bovine serum albumin (138), rabbit muscle pyruvate kinase (176), apoconcanavalin A (175), *B. subtilis* apo-α-amylase (163), Myeloma Protein MOPC 315 (194f), and to the Chromatophore membrane in *Chromatium* (145a).

In the ensuing discussion it is assumed that the contact terms in Eqs. (11) and (12) may be ignored, and that conditions of fast exchange abound in the various systems. These specifications have been verified by temperature variation studies of relaxation rates in all cases but one [vide supra, (134), (135)]. Thus $1/T_{xM}$ is a function of τ_c, and if this correlation time is known the simplified

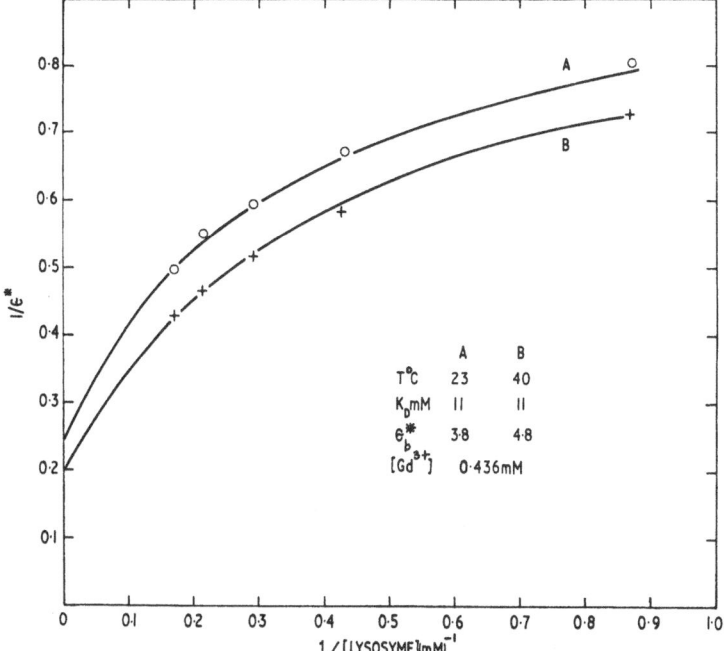

Fig. 5. Titration of lysozyme with 436 μM Gd(III) at 32 °C and 40 °C. The continuous lines are computed for the parameters indicated in the figure. [Reproduced by permission from *Dwek, R. A., Richards, R. E., Morallee, K. G., Nieboer, E., Williams, R. J. P., Xavier, A. V.*: European J. Biochem. **21**, 204 (1971)]

Solomon-Bloembergen Eqs. [cf. (11) and (12)] can be solved for r, the resonating nucleus-metal ion separation. Conversely, estimated values of r (from X-ray data) and τ_c (vide infra) allow the computation of $1/T_{x\text{M}}$, which is useful in water proton relaxation enhancement studies in the evaluation of q, the primary hydration number of the metal ion. In this instance, p_L in Eq. (15) corresponds to $C_\text{M} q/55.5$, where C_M is the total concentration of the paramagnetic ion. The quantity $q/T_{x\text{M}}$ can be determined experimentally [Eq. (15)]; division of this quantity by the calculated value of $1/T_{x\text{M}}$ yields a value for q. *Reuben* (138) determined that for Gd^{+3} q was equal to 9. The value of τ_c for the aquo-complex of Gd^{+3} was determined independently by *Dwek et al.* (135) and *Reuben* (138) from the temperature and frequency dependencies of the experimental relaxation rates $1/T_1$ and $1/T_2$. It could further be identified with the rotational tumbling time of the aquo complex ($\tau_c = \tau_\text{R} = 7 \times 10^{-11}$ sec at 300 °K).

When Gd(III) binds to a macromolecule its tumbling time is slowed down and thus $\tau_c^* > \tau_c$ and $1/T_{x\text{M}}^* > 1/T_{x\text{M}}$. [The asterisk simply indicates the presence of the macromolecule; Eq. (15) applies equally to the starred quantities.] Substitution of expression (11) or (12) into relationship (17) for both $1/T_{x\text{M}}^*$ and $1/T_{x\text{M}}$, and inserting an arbitrary value of q^*, it is possible to solve for τ_c^* since all other parameters are either known (e.g., τ_c, ω_I, ω_S) or cancel out (r). Besides the choice of q^*, errors in this evaluation could arise from the anisotropic rotation of the paramagnetic ion and its waters of hydration about the metal-macromolecule linkage. Such process is expected to reduce the dipolar interaction between the water molecule and the central paramagnetic ion (134). Morallee et al. (7) used this approach in evalauting τ_c for the Gd^{+3}-lysozyme complex. The Gd^{+3} perturbations of the line widths (line-width at half-height $= 1/\pi T_2$) corresponding to protons on the substrate bound to the Gd^{+3} lysozyme complex where subsequently evaluated. They assumed that the previously evaluated τ_c^* also modulated the magnetic interaction for these substrate protons. From the experimentally evaluated $1/T_{2M}^*$ quantity [Eq. (15)], r was calculated from the simplified form of (12). The metal-proton distance determined in this manner agreed favourable with those estimated from the three-dimensional structure of this enzyme (vide supra): r (acetamide) $= 6.7$ Å (6.8) and r (glycosidic) $= 5.6$ Å (4.6), where the solid state separations are those quoted in the brackets.

Another procedure for evaluating τ_c is by solving the ratio $(1/T_{1M})/(1/T_{2M})$ corresponding to the same proton for this parameter. High resolution NMR with Fourier transform capability is required if this is to be done for protons of a protein or for protons located on an enzyme-bound substrate. Dwek et al. (205) reinvestigated the lysozyme system just described using this approach. Similar results were again obtained. Several authors have emphasized the uncertainties and difficulties in these approaches to the measurement of τ_c^* (205, 206). However, it should be noted that the occurrence of r to the sixth power in expressions (11) and (12) serves to render r calculations rather insensitive to even large errors. For example, a factor of 10 in τ_c changes the absolute distances by a factor of 1.46; of course, errors in τ_c^* are avoided in relative distance calculations.

Nieboer et al. (129) have recently employed the ratio method in broadening studies of the C2 histidine resonances of Staphylococcal nuclease. This enzyme requires calcium to activate it and is known to have four histidine amino acid residues in its primary sequence. All four C2 imidazole proton resonances are observed in the aromatic region of the protein spectrum (see Fig. 6). The need to check for diamagnetic effects with La^{+3} is vividly illustrated in this figure. La^{+3} would appear to be spatially near enough to the H-2 imidazole moiety to perturb the corresponding C-2 proton nucleus. Absolute assignments of these four resonances were made by comparing the experimental Gd^{+3}-histidine distances with estimated values based on the 3D-structure of a Ca^{+2}-nuclease-substrate complex (51, 52). The RE ions are known to compete with Ca^{+2} for the same binding site from competitive inhibition studies (98, 207). Relevant data is reproduced in Table 4. The magnitude of the binding constant at pH 5.25 $(2.8 \times 10^2$ M$^{-1})$ indicates considerably weaker binding than at pH 7 $(\sim 10^5$ M$^{-1})$. Water proton relaxation data are consistent with the loss of one water molecule on complex formation at pH 5.25, and thus complex formation with a single carboxylate group is therefore implied. In contrast, the Ca(II) in the nuclease-inhibitor complex is co-ordinated by an approximately square array of carboxylate groups. This condition probably persists in solution at pH 7. In spite of this difference in denticity between the complex at 5.25 and that in the crystal, as well as the absence of the inhibitor in the work at pH 5.25, the experimental metal-to-histidine distances are in remarkable agreement with the X-ray values. The largest deviation between these two sets is for His-8, and this may be significant. The shorter

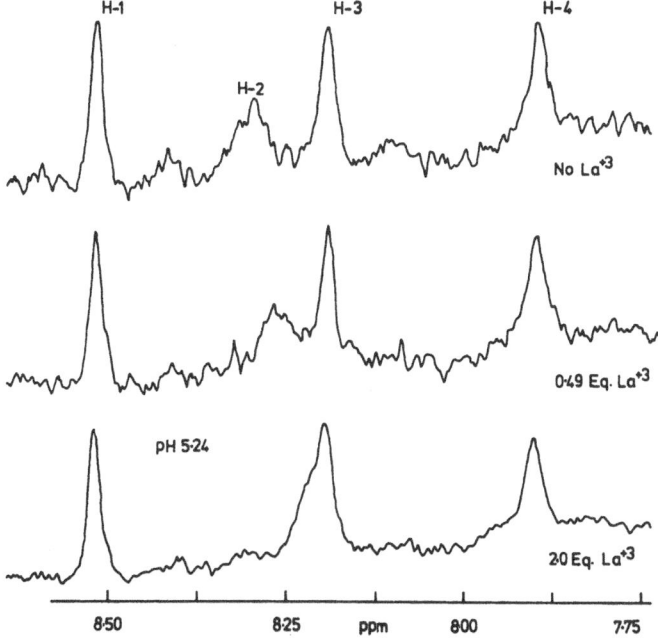

Fig. 6. The effect of La^{+3} on the C2 imidazole proton region of the 220 MHz spectrum of Staphylococcal nuclease at pH 5.25 (Nuclease was 2.5 mM, chemical shifts refer to an external TMS reference, and spectra correspond to 20 time-averaged scans; Eq = equivalents)

distance may reflect movement in solution of the N-terminal end of the peptide chain, since in the crystal residues 1 to 5 cannot be located accurately. Nevertheless, this study confirms the notion that a protein in solution retains the gross features of its solid state 3D structure.

Table 4. ^{1}H-Gadolinium distances calculated from relaxation enhancement data for the C2 imidazole protons of staphylococcal nuclease[a])

	H-1	H-2	H-3	H-4
$\bar{r}(\text{Å})_{\text{exp}}$ [b])	19.2	8.8	16.0	15.2
$\bar{r}(\text{Å})_{\text{x-ray}}$ [c])	26.2	9.0	19.6	17.3
Assignment	His-8	His-46	His-124	His-121

[a]) All spectra were recorded at pH 5.25 and 0.1 M NaCl; a typical spectrum of nuclease is given in Fig. 6; $\tau_c^* = 2.6 \times 10^{-9}$ sec at 22 °C, the experimental temperature.

[b]) Determined from the NMR data.

[c]) Measured distances from Ca^{+2} ion to closest atom of imidazole side chains from X-ray data (51, 52) of nuclease-Ca^{+2}-thymidine 3',5'-diphosphate complex (crystals were prepared at pH 8.2) and based on the assignment indicated.

Furie et al. *(208)* have extended this study of Staphylococcal nuclease by measuring the relaxation enhancement of ^1H and ^{31}P nuclei of the inhibitor 3',5'-thymidine diphosphate when bound to the Gd^{+3}-nuclease complex at pH 6.9. In general, the experimental distances correspond closely to estimates obtained from the X-ray data of the Ca^{+2}-nuclease-inhibitor ternary complex. The experimental separations were then employed in solving for nucleotide geometries relative to the metal position. A geometry similar to the structural arrangement found in the solid state was one of the possible solutions to this computer modeling process. Thus for Staphylococcal nuclease, the NMR and X-ray methods yield compatible high resolution information about the structure of the active site. However, differences of uncertain significance exist between the two structures. This is not altogether surprising since earlier work *(98)* had indicated a marginal difference in the mode or position of binding of RE ions relative to Ca^{+2}. In contrast to Ca^{+2}, the lanthanide ions at pH 7 induced perturbations of tyrosyl electronic transitions and rendered the lanthanide nuclease complex resistant to trypsin digestion. This difference was also reflected in the inhibitory action of the RE ions in the Ca^{+2} generated nuclease activity. It is of interest that the X-ray diffraction studies *(51, 52)* of crystalline nuclease showed that the centers of the aromatic rings of tyrosines 115 and 85 are 13.3 Å and 12.4 Å, respectively, from the bound Ca^{+2} in the active site. Direct interaction of these tyrosines with the RE metal ions is discounted because of this, and because the observed spectral perturbation was not due to a distinguishable alteration in the pK of the tyrosyl-115-hydroxyl group. It seems significant that no fluorescence is observed for nuclease with Tb^{+3}, nor with Eu^{+3}. This supports our previous contention that Tb^{+3} fluorescence depends on the location of an aromatic donor group at the periphery of the metal co-ordination spheres.

Absolute distances relative to the Gd^{+3} reporter site to protons on an enzyme bound substrate have also been estimated from relaxation enhancement studies for muscle pyruvate kinase *(176)*, bovine trypsin *(194p)* and concanavalin A *(194q)*.

v) Chemical Shift Probe Studies

The EuEDTA serves as an informative model for europium shift studies. Typical spectra *(96)* of this complex are depicted in Fig. 7 and 8. The AB pattern observed at elevated temperatures is assigned to the methylenic protons and the singlet to the ethylenic protons. All signals broaden drastically as the temperature is lowered. (A medium of 3 M LiCl circumvents the inherent solubility problem of the system and appears not to affect the observations.) Around 19 °C additional peaks appear in the spectrum whose intensities grow at the expense of the original peaks as the temperature is lowered to -10 °C. Consequently, the existence of two species in this system is confirmed. Recent work in the author's laboratory on the temperature dependance of the Li(La EDTA) spectrum clearly indicates the presence of an unbound carboxylate group. At temperatures ≤ 10 °C a sharp singlet splits out from the methylenic resonance (singlet) which does not broaden as the temperature is lowered; concomitantly the remaining methylenic peak, as well as the ethylenic peak, becomes extremely broad at the lower temperatures

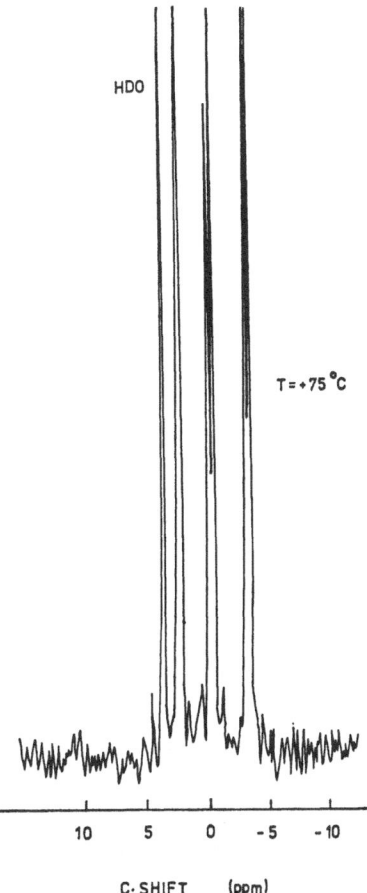

Fig. 7. The 60 MHz spectrum of 0.1 M NaEuEDTA at 75 °C. Chemical shifts are relative to an external DSS reference

although no additional resonances appear. This line-broadening is consistent with the slowing down of both ethylenic and methylenic ring flipping processes that do not require La-O bond breakage. The conclusion of *Kostromina et al.* (*89, 90*) that EDTA in the Eu^{+3} low temperature species (likely the only species for La EDTA$^-$ at normal temperatures) has a lower denticity appears to be confirmed. Addition of excess EDTA to 0.1 M NaEuEDTA caused observable shifts in all the NaEuEDTA peaks (*96*), and thus the reported existence of a 2:1 ligand-metal species (*93*) was substantiated (log $K_2 = 2.65$ at 45 °C and pD $= 7.5$). The second ligand remains singly protonated on binding. Water relaxation enhancement studies (*96*) were consistent with the assignment of $x = 4$ in Eq. (6), with $j = 1$ of course. Two waters of hydration appear to remain on the europium in the 2:1 species.

The inherent warning in this data for the EuEDTA complex is that in tightly bound ligand systems intramolecular isomerization processes, isomer exchange,

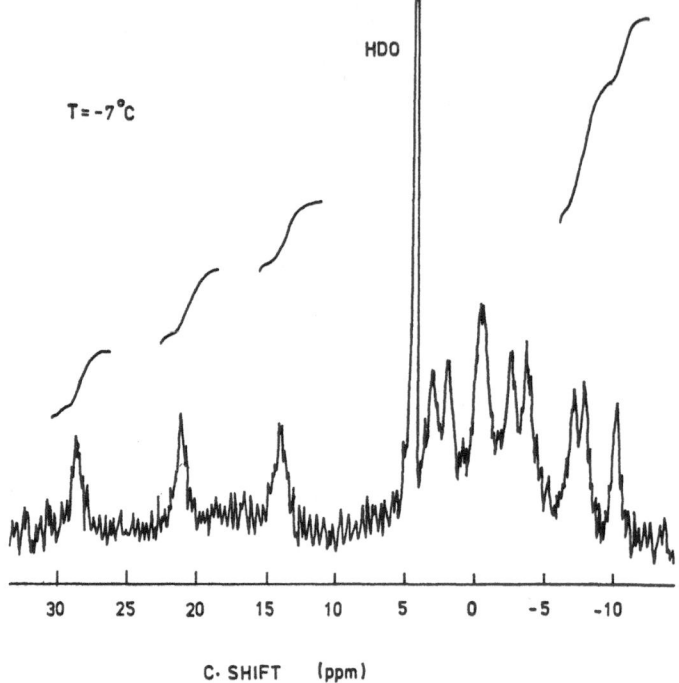

Fig. 8. The 60 MHz spectrum of 0.1 M NaEuEDTA in 3M LiCl at −7 °C. Chemical shifts are relative to an external DSS reference

and secondary ligand exchange may dominate the observed line-widths. The ability to study such processes by NMR is of course a definite advantage.

Podolski et al. (*128, 209*) have applied chemical shift studies to the primary sequence elucidation of simple peptides. At 60 MHz detectable shifts could be observed for protons up to the 5th residue. Peptides of more than four amino acid residues were found to be soluble in 3M sodium perchlorate. The pH of the study was chosen as 3.7 so that the amino group of the N-terminal was protonated; the RE ion was found not to compete with the proton for this binding site, in agreement with other workers (*203, 210*). Gd^{+3} relaxation studies confirmed that the magnitude of the observed shifts was proportional to the reciprocal distance between the metal at the C-terminal and the ^1H nucleus observed. Double reciprocal mole-ratio plots $1/\Delta\nu$ *versus* C_L/C_M, were found to be linear over a wide range of metal concentration for titrations performed at constant ligand concentration. Two examples are given in Fig. 9 and 10. [The nomenclature employed to identify ^1H nuclei, $CH_{\alpha,x}$ or $CH_{\beta,x}$, indicates the position of a nucleus in a given residue (α, or β) and the residue number (x) from the C-terminal.] The intercept of such plots corresponds to the reciprocal of the induced shift of the fully formed complex — the limiting shift. The slope may be employed in the calculation of $\bar{n}$ at all points in a titration. The quantity $\bar{n}$ represents the average number of ligands bound to the metal ion (*211*). This permits the determination of the formation function (*211*), from which all step-wise formation constants may be evaluated.

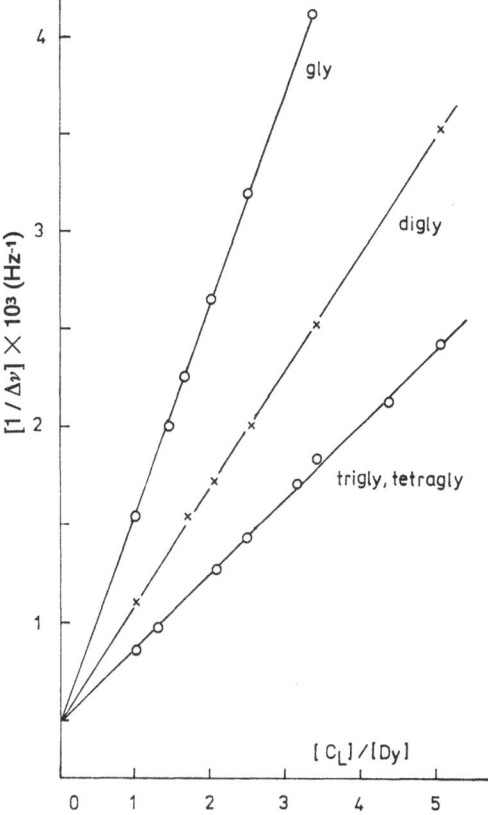

Fig. 9. Double reciprocal mole-ratio plots for the induced 60-MHz $CH_{\alpha,1}$ proton shifts of glycine, diglycine, triglycine and tetraglycine

No prejudice need to be introduced concerning the stoichiometry of the complexes as is the case in another NMR method (*202*). The common intercept for the $CH_{\alpha,1}$ protons of the glycines in Fig. 9 implies that the geometry of the complexes are identical. The different slopes show that the binding in the case of glycine is the weakest, and this reflects the short distance between the positively charged amino group and the metal ion. In conclusion, the data in Table 5 clearly indicates that the induced shift corresponding to the fully formed complex is a good diagnostic indicator for the position of a residue in a peptide chain made up of glycine residues, or a combination of glycine and alanine residues.

In Fig. 11 chemical shifts perturbations for the histidine C2 resonances of Staphylococcal nuclease are plotted as a function of the metal-to-ligand ratio. Unlike in the peptide studies, the instrumentation used for this protein (220 MHz) precluded the use of an internal standard. Therefore, the observed shift is the sum of the bulk susceptibility and any specific dipolar shifts. However, the unidentified resonance R-5 was found to shift at a rate equal to the bulk susceptibility perturbations. Absolute shifts could therefore be evaluated relative to R-5. The double

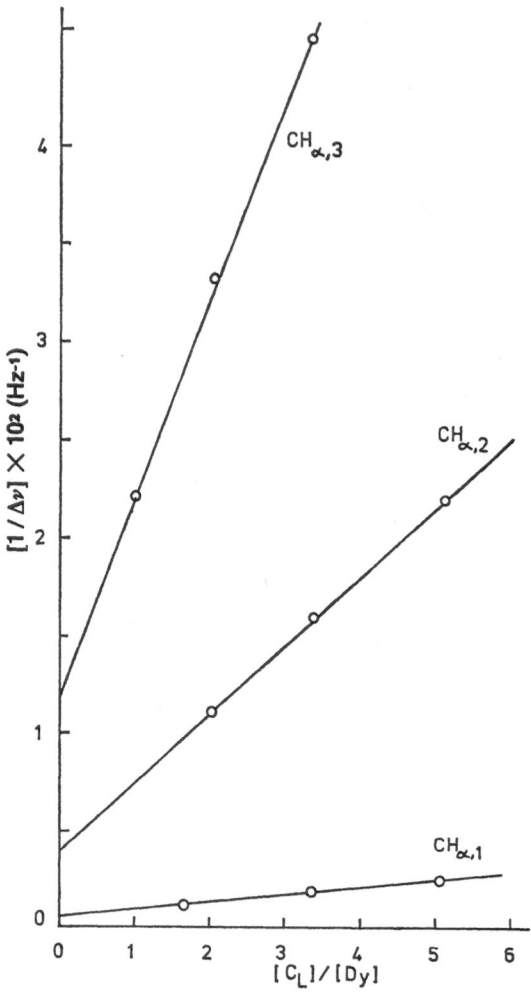

Fig. 10. Double reciprocal mole-ratio plots for the induced 60-MHz $CH_{\alpha,1}$, $CH_{\alpha,2}$, and $CH_{\alpha,3}$ proton shifts of triglycine

Table 5. Induced limiting shifts for ^1H nuclei of peptides complexed to Dy^{+3} at 60 MHz

Limiting Shift (Hz)[a, b, c]

$CH_{\alpha,1}$	$CH_{\alpha,2}$	$CH_{\alpha,3}$	$CH_{\alpha,4}$	$CH_{\alpha,5}$	$CH_{\beta,1}$	$CH_{\beta,2}$[d]	$CH_{\beta,3}$
2000	250	83	33	10	1250	100	50
±100	±15	±5	±3	±2	±50	±5	±3

[a] Evaluated from double reciprocal plots (cf., Fig. 9 and 10).

[b] See text for definition of symbols.

[c] Peptides examined were: gly, digly, trigly, tetragly, pentagly, glyalagly, alaglygly, gly-alaala, glyglyala, alaalagly, ala, triala (the first residue at the left of each peptide name is the N-terminal one).

[d] Consistently a value of 67 ±5 was observed when the C-terminal residue was an alanine. This may imply steric inhibition to rotation for this combination.

reciprocal plots, $1/\Delta\nu$ versus C_L/C_M, were again found to be applicable. Shift ratios, $\Delta\nu_3/\Delta\nu_x$, where the subscript identifies the histidine resonance in question, were found (129) to be independent of the metal ion (effective axial symmetry!), and their cube root were equal in magnitude to ratios calculated from the experimental absolute distances in Table 4 except for H-1. Consequently, the term $(3\cos^2\Theta-1)$ contributes to the ratio $\Delta\nu_3/\Delta\nu_1$.

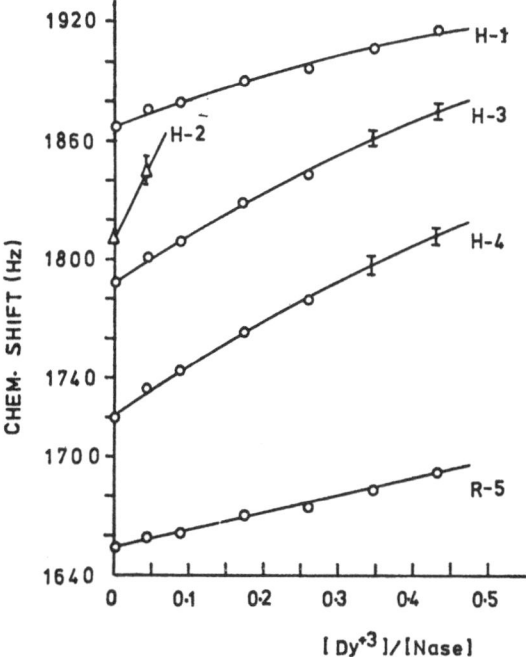

Fig. 11. Mole ratio plot for the observed shifts at 220 MHz corresponding to the C2 imdizole protons of Staphylococcal nuclease. R-5 is an arbitrary selected reference peak present in the aromatic region. Chemical shifts are relative to TMS (external)

Finally, Bystrov et al. (212) showed that the proton NMR signals of the $-\overset{+}{N}(CH_3)_3$ groups of lecithin molecules forming the internal and external surfaces of bilayer vesicles can be differentiated using the paramagnetic cation Eu^{+3} in combination with Mn^{+2} or Gd^{+3} (213). (A suspension of vesicles is generated by sonicating a dispersion of egg white lecithin in water; lecithin is a phospholipid.) Signals arising from the choline methyl protons are split into two peaks under the influence of Eu^{+3}, and the shifted area is also that broadened by Mn^{+2} or Gd^{+3}. This result was interpreted as indicating that the paramagnetic cations can only interact with the phosphate groups of lecithin molecules on the outer surface of the vesicles. Because the bilayer is semipermeable, the metal ions can not complex with the phosphate groups on the "inner" surface of the vesicles. Ratios of integrated areas of shifted and unshifted peaks gave the relative number of "external" and "internal" $-N^+(CH_3)_3$ groups, and were found to be in good agreement with the relative surface areas calculated from the dimensions of the spherical vesicles (214). Very recently, *Huang et al.* (215) have exploited this specific chemical shift perturbation in a study of the effect of incorporating cholesterol in the lipid phase of these bilayers. When the cholesterol content exceeded 25 mol % a significant and rather abrupt increase in the area ratio external/internal was observed. These results were interpreted as due to the asymmetric distribution of the cholesterol in favour of the inner layer. Steric hindrance to packing of the large unsaturated molecules of hen egg phosphatidyl-choline was expected to be most severe at the "inner" surface because of its low radius of curvature. The preferential displacement of these molecules by the smaller cholesterol molecules was thus accounted for. Preference for the "inner" surface was not detected for the vesicles of saturated L-α-dipalmitoyl lecithin. Extension of these conclusions suggested that lipid vesicles with low radii of curvature may be especially appropriate models for the "active site" regions of membranes. It is generally known that a membrane is not uniform, but that for each of its many functions only specialized areas are effective. These areas often occur in highly folded regions of a membrane and thus possess low radii of curvature.

Relaxation (Mn^{+2}) and shift studies (Eu^{+3}) of both ^1H and ^{31}P nuclei in bilayers prepared from equimolar quantities of the anionic phospholipid phosphatidyl-glycerol and the zwitterion phosphatidylcholine, have provided further evidence for the chemical asymmetry of vesicles (216). The outer surface was found to contain twice as many molecules of the glycerol derivative than the choline phospholipid, and these molecules were not spatially segregated into "patches". These results support the view that the electrostatic repulsion between the negatively charged phosphatidylglycerol molecules would be most severe at the inner surface because of its small radius of curvature, and that the observed asymmetric distribution is electrostatically favoured. The metal ions were also found to bind preferentially to the anionic phospholipid molecules. Similarly, information has also been obtained for lipid-protein interactions (217). Since no unshifted component could be observed for the ^{31}P resonance in the presence of Eu^{+3}, it was concluded that essentially all the phospholipid phosphorous was located at the outer surface of the high-density lipoprotein particles obtained from human serum.

In summary, and as predicted by *Williams* (218), lanthanide induced relaxation and shift perturbations would appear to be ideally suited for studying the chemical asymmetry of membranes, especially that due to diluents such as cholesterol, proteins, anesthetics, and electron transport intermediates. Quantitative spatial information is in principle possible since *Andrews et al.* (219) have shown that for ^1H nuclei in the phosphatidylcholine system the dipolar mechanism predominates for the paramagnetic interaction with the lanthanide ions. Since so many biological functions of a living cell occur at lipid interfaces, the application of chemical shift and relaxation probes to heterogeneous systems is no doubt a challenge of considerable importance and urgency.

vi) Conformational Studies

For a RE-ligand complex of axial symmetry a shift ratio, S_i, is readily defined for two nuclei on the same ligand molecule.

$$S_i = \Delta_{\nu_i}^* / \Delta_{\nu_0}^* = (G_i(\Theta)/G_0(\Theta))\, r_0^3/r_i^3 \tag{19}$$

$G_i(\Theta) = 3\cos^2\Theta - 1$ as in Eq. (9), and the subscript 0 designates one of the nuclei as an arbitrary reference. Similarly, a broadening ratio, B_i, may be defined for the same nuclei

$$B_i = (T_{xM}^{-1})_i/(T_{xM}^{-1})_0 = r_0^6/r_i^6 \tag{20}$$

The empirical parameters S_i and B_i contain geometric and distance information. *Barry et al.* (*127, 220*) evaluated S_i and B_i for ^{1}H nuclei of mononucleotides under acidic conditions (pD = 2) in the presence of the shift probes Eu^{+3} and Ho^{+3} and the broadening agent Gd^{+3}. They then proceeded to evaluate the most probable conformation for these molecules (*e.g.*, AMP) in solution in the following manner. Their approach is simple but requires computer facilities with graphic display capability. Molecular coordinates (from X-ray data, or from known bond distances and angles) are fed into the computer. It is then given Van der Waals radii and one S_i value, and instructed to search for agreement. The metal ion is then placed (near the monophosphate anion in the present case) such that acceptable solutions are possible. Additional S_i values and B_i values are then provided as filters and the best solution is searched out. For example, 12 members of a distinct geometric family of conformers gave the best fit for AMP and these were accepted as representing the most stable conformations in solution. Improved resolution in these studies has been achieved (*220, 221*) by using a combination of ^{13}C and ^{1}H data, and employing longitudinal nuclear relaxation rates (T_1 measurements) instead of the less reliable transverse rates (T_2 measurements). Availability of Fourier Transform NMR facilities has made this feasible. ^{31}P shift and relaxation data may also be employed, although it is necessary to separate the contact and dipolar contributions to the induced shift (*222*). Similar conformations for adenosine 5'-monophosphate and cytidine 5'-monophosphate were found in solutions of pH 2 and 7.5, and neither did temperature have much effect on these observations (*223*). Studies in basic solution were made possible by using the lanthanide complexes of EDTA as shift and broadening agents, as these are not hydrolyzed as readily as the aquo ions. The independence of the observed proton shift ratios from the specific lanthanoid-EDTA complex used confirmed that effective axial symmetry existed for the resulting nucleotide complexes as was the case (*127, 162, 223*) for the aquo ions (except Tm^{+3}). The quantitative determination of solution conformations of dinucleoside phosphates were affected by the same techniques (*162* and references therein), as well as those of cyclic nucleotides (*221, 224*). It is also possible to extent these methods to polar organic solvents such as DMSO and methanol (*162*). In non-polar solvents, the appropriate β-diketonates, *e.g.*, Ln(thd)$_3$, may be employed as shift and relaxation (Gd^{+3}) reagents. *Barry et al.* (*225*) found that both relaxation and shift data (^{13}C and ^{1}H) were essential in the evaluation with confidence of the structure of the $Ln(thd)_3 \cdot$ cholesterol adduct in $CDCl_3$. Variation of solvent might allow insight into some of the factors that determine conformation. Extrapolation to *in vivo* conformations of these biological molecules might then be attempted with fewer assumptions.

Dr. *R. J. P. Williams'* group at Oxford are developing even more versatile methods to solution structural studies. Combination of high frequency (*e.g.* 270 MHz), Fourier transform, deconvolution (manipulating spectrometer response in a certain mathematical manner), and a judicious application of shift and broadening agents has made it possible to obtain very highly resolved proton spectra of lysozyme (*191*). As indicated earlier in this article, Gd^{+3} is known to bind near the active site of lysozyme, and will perturb the resonances of protons within its sphere of influence. NMR difference spectra between free lysozyme and the Gd^{+3}-lysozyme complex consist only of the resonances whose intensities have been reduced to zero by dipolar broadening. (The degree and the range of broadening may be controlled by the amount of Gd^{+3}-lysozyme complex since conditions of fast exchange persist in this system; a little bound Gd^{+3} will only broaden resonances of protons at close range, but as the fraction of complex increases protons further away are affected.) Attention may thus be focused on the immediate environment around the metal. Shift reagents may be used similarly to shift peaks away so that they appear in the difference NMR spectrum. Assignments of resonances in the difference spectrum can then be made with the aid of double resonance techniques and by reference to the known 3D solid state structure. Once all the assignments have been made, S_i and B_i values are to be determined and used in the quantitative determination of the effective solution protein structure as described earlier for the mononucleotides. Preliminary results (*191*) have confirmed the feasibility of this approach, and thus a new and exciting era has been ushered in for structural and conformational studies of macromolecules in solution.

vii) Concluding Remarks

The lanthanide ions have traditionally been considered as a group of ions with indistinguishable chemistry, which by transition metal standards is non-spectacular. It is fortunate that their simple type of ionic chemistry plays such a dominant role in the chemistry of living systems. If this were not so, the application of the RE ions as structural probes would be less universal. The almost instant road to prominence as reporter groups has revealed an individuality for each member of the lanthanoid series that is not only fascinating, but also of extreme practical value. There is no doubt that these ions have already secured a permanent function in solving biological problems, regardless of future developments.

Addendum:

Additional information is available on the following subjects: binding of Gd^{+3} and Tb^{+3} to potentially bidentate monocarboxylates (*226*); structure and denticity of small peptides in solution (*227*); water-proton relaxation enhancement (*228*) and mapping studies of lysozyme (*229*); NMR investigation of yeast phosphoglycerate kinase (*230*) and its inhibition by lanthanide-ATP complexes (*231*);

spin-label and lanthanide binding studies of glyceraldehyde-3-phosphate dehydrogenase *(232)*; use of Pr^{+3} in the assignment of aromatic amino acid PMR resonances of horse ferricytochrome C *(233)*; and primary sequencing of peptides *(234, 235)*.

Acknowledgements. I would like to express my appreciation to Dr. *R. J. P. Williams* of the Oxford Enzyme Group, University of Oxford, for introducing me to the subject matter of this review, and I thank him for helpful comments on the manuscript.

References

1. *Cohn, M.:* Quart. Rev. Biophys. *3*, 61 (1970).
2. *Latt, S. A., Vallee, B. L.:* Biochemistry *10*, 4254 (1971).
3. *Lee, G. A.:* Coord. Chem. Rev. *8*, 290 (1972).
4. *Vallee, B. L., Williams, R. J. P.:* Chem. Brit. *4*, 397 (1968).
5. *Birnbaum, E. R., Gomez, J. E., Darnall, D. W.:* J. Am. Chem. Soc. *92*, 5287 (1970).
6. *Williams, R. J. P.:* Quart. Rev. Chem. Soc. (London) *XXIV*, 231 (1970).
7. *Morallee, K. G., Nieboer, E., Rossotti, F. J. C., Williams, R. J. P., Xavier, A. V.:* Chem. Commun. *1970*, 1132.
8. *Hinckley, J.:* Am. Chem. Soc. *91*, 5160 (1969).
9. *Horrocks, W., DeW., Jr., Sipe, J. P., III:* J. Am. Chem. Soc. *93*, 6800 (1971).
10. *Reuben, J.:* Progr. NMR Spectrosc. *9* (1), 1 (1973).
11. *Sievers, R. E.:* Nuclear magnetic resonance shift reagents. New York: Academic Press 1973.
12. *Mayo, B. C.:* Chem. Soc. Rev. *2* (1), 49 (1973).
13. *Cockerill, A. F., Davies, G. L. O., Harden, R. C., Rackman, D. M.:* Chem. Rev. *73*, 553 (1973).
14. *Lind, M. D., Lee, B., Hoard, J. L.:* J. Am. Chem. Soc. *87*, 1611 (1965).
15. *Muetterties, E. L., Wright, C. M.:* Quart. Rev. Chem. Soc. (London) *XXI*, 109 (1967).
16. *Al-Karaghauli, A. R., Wood, J. S.:* J. Am. Chem. Soc. *90*, 6548 (1968).
17. *Al-Karaghauli, A. R., Wood, J. S.:* Inorg. Chem. *11*, 2293 (1972).
18. *Forsberg, J. H.:* Coord. Chem. Rev. *10*, 195 (1973).
19. *Hoard, J. L., Lee, B., Lind, M. D.:* J. Am. Chem. Soc. *87*, 1612 (1965).
20. *Stezowski, J. J., Countryman, R., Hoard, J. L.:* Inorg. Chem. *12*, 1749 (1973).
21. *Wells, A. F.:* Structural inorganic chemistry (3rd ed.). Oxford: Oxford Clarendon Press 1962.
22. *Grenthe, I.:* Acta Chem. Scand. *23*, 1752 (1969).
23. *Grenthe, I.:* Acta Chem. Scand. *26*, 1479 (1972).
24. *Grenthe, I.:* Acta Chem. Scand. *25*, 3347 (1971).
25. *Albertson, J.:* Acta Chem. Scand. *24*, 3527 (1970).
26. *Albertson, J.:* Acta Chem. Scand. *22*, 1563 (1968).
27. *Frost, G. H., Hurt, F. A., Heath, C., Hursthouse, M. B.:* Chem. Commun. *1969*, 1421.
28. *Albertson, J.:* Acta Chem. Scand. *26*, 985, 1005, 1023 (1972).
29. *Horrocks, W., DeW., Sipe, J. P., III, Luber, J. R.:* J. Am. Chem. Soc. *93*, 5258 (1971).
30. *Cramer, R. E., Seff, K.:* Chem. Commun. *1972*, 400.
31. *Phillips, I., II, Sands, D. E., Wagner, W. F.:* Inorg. Chem. *7*, 2295 (1968).
32. *Shineman, R. S., King, A. J.:* Acta Cryst. *4*, 67 (1951).
33. *MacLennan, G., Beevers, C. A.:* Acta Cryst. *9*, 187 (1956).
34. *Dickens, B., Prince, E., Schroeder, L. W., Brown, W. E.:* Acta Cryst. B *29*, 2057 (1973)·
35. *Uchtman, V. A., Oertel, R. P.:* Proc. XIVth Int. Conf. Coord. Chem. 182 (1972); J. Am. Chem. Soc. *95*, 1802 (1973).
36. *Bugg, C. E., Cook, J. W.:* J. C. S. Chem. Commun. 727 (1972).
37. *Cook, W. J., Bugg, C. E.:* Acta Cryst. B *29*, 215 (1973).
38. *DeVilliers, J. P. R., Boeyens, J. C. A.:* Acta Cryst. B *27*, 692 (1971).
39. *Erasmus, C. S., Boeyens, J. C. A.:* Acta Cryst. *826*, 1843 (1970).
40. *Erasmus, C. S., Boeyens, J. C. A.:* J. Cryst. Mol. Struct. *1*, 83 (1971).

41. *Schuchart Wasson, S. J., Sands, D. E., Wagner, W. F.:* Inorg. Chem. *12*, 187 (1973).
42. *Dyer, D. S., Cunningham, J. A., Brooks, J. J., Sievers, R. E., Rondeau, R. E.:* In: Nuclear magnetic resonance shift reagents, p. 21 (*Sievers, R. E.*, ed.). New York: Academic Press 1973.
43. *Bouloux, J. C., Galy, J.:* Acta Cryst. B *29*, 269 (1973).
44. *Van der Helm, D., Willoughby, T. V.:* Acta Cryst. B *25*, 2317 (1969).
45. *Trueblood, K. N., Horn, P., Luzatti, V.:* Acta Cryst. *14*, 965 (1961).
46. *Whitlow, S. H.:* Acta Cryst. B *28*, 1914 (1972).
47. *DeVilliers, J. P. R., Boeyens, J. C. A.:* Acta Cryst. B *28*, 2335 (1972).
48. *Busing, W. R., Levy, H. A.:* J. Chem. Phys. *26*, 563 (1957).
49. *Kretsinger, R. H., Nockolds, C. E.:* J. Biol. Chem. *248*, 3313 (1973).
50. *Edelman, G. M., Cunningham, B. A., Reeke, G. M., Jr., Becker, J. W., Waxdal, M. J., Wang, J. L.:* Proc. Natl. Acad. Sci. (U.S. *69*, 2580 (1972).
51. *Arnone, A., Bier, C. J., Cotton, F. A., Day, V. W., Hazen, E. E., Jr., Richardson, D. C., Richardson, J. S., Yonath, A.:* J. Biol. Chem. *246*, 2302 (1971).
52. *Cotton, F. A., Bier, J. C., Day, V. W., Hazan, E. E., Larsen, S.:* Cold Spring Harbor Symp. Quant. Biol. *36*, 243 (1971).
53. *Matthews, B. W., Colman, P. M., Jansonius, J. N., Titani, K., Walsh, K. A., 2nd, Neurath, H.:* Nature (London), New Biol. *238*, 41 (1972)
54. *Colman, P. M., Weaver, L. H., Matthews, B. W.:* Biochem. Biophys. Res. Commun. *46*, 1999 (1972); Biochemistry *13*, 1719 (1974).
55. *Martin, J. L., Thompson, L. C., Radanovich, L. J., Glick, M. D.:* J. Am. Chem. Soc. *90*, 4493 (1968).
56. *Moeller, T.:* In: Inorganic chemistry series one, Vol. 7, p. 275 (*Bagnall, K. W.*, ed.). Baltimore: Univ. Park Press 1972.
57. *Nieboer, E., McBryde, W. A. E.:* Can. J. Chem. *51*, 2511 (1973).
58. *Williams, R. J. P., Hale, J. D.:* Struct. Bonding *1*, 249 (1966).
59. *Pearson, R. G.:* J. Chem. Educ. *45*, 581 (1968).
60. *Klopman, G.:* J. Am. Chem. Soc. *90*, 223 (1968).
61. *Pearson, R. G.:* J. Chem. Educ. *45*, 643 (1968).
62a.*Jørgenson, C. K.:* Struct. Bonding, *3*, 106 (1967).
62b.*Choca, M., Ferraro, J. R., Nakamoto, K.:* Coord. Chem. Rev. *12*, 295 (1974).
63. *Cotton, F. A., Wilkinson, G.:* Advanced inorganic chemistry, Chapt. 27 (3rd Ed.) New York: Interscience 1972.
64. *Kolat, R. S., Powell, J. E.:* Inorg. Chem. *1*, 293 (1962).
65. *Grenthe, I., Tobiasson, I.:* Acta Chem. Scand. *17*, 2101 (1963).
66. *Anderegg, G., Wenk, F.:* Helv. Chim. Acta *54*, 216 (1971).
67. *Dadgar, A., Choppin, G. R.:* J. Coord. Chem. *1*, 179 (1971).
68. *Jørgensen, D. K.:* Struct. Bonding *13*, 199 (1973).
69. *Karraker, D. G.:* J. Chem. Educ. *47*, 424 (1970).
70. *Moeller, T., Martin, D. F., Thompson, L. C., Ferrus, R., Feistel, G. R., Randall, W. J.:* Chem. Rev. *65*, 1 (1965).
71. *Phillips, C. S. G., Williams, R. J. P.:* Inorganic chemistry Vol. 2, Chapt. 21. Oxford: Oxford University Press 1965.
72. *Millero, F., J.:* Chem. Rev. *71*, 147 (1971).
73. *Johnson, D. A.:* J. Chem. Soc. (A), 1525 (1969).
74. *Spedding, F. H., Pikal, M. J., Ayers, B. O.:* J. Phys. Chem. *70*, 2440 (1966); Proceedings of the Eleventh Rare Earth Conference, II, p. 909, 919. Traverse City Mich.: U.S. Dept. of Commerce, Springfield, Va. October 1974.
75. *Grenthe, I., Hessler, G., Ots, H.:* Acta Chem. Scand. *27*, 2543 (1973).
76. *Reuben, J., Fiat, D.:* J. Chem. Phys. *51*, 4909, 4918, (1969).
77. *Bockris, J. O'M., Saluja, P. P. S.:* J. Phys. Chem. *76*, 2298 (1972).
78. *Hepler, L.:* J. Phys. Chem. *69*, 965 (1965).
79. *Choppin, G. R.:* Pure Appl. Chem. *27*, 23 (1971).
80. *Choppin, G. R., Graffeo, A. J.:* Inorg. Chem. *4*, 1254 (1965).
81. *Grenthe, I., Ots, H.:* Acta Chem. Scand. *26*, 1217, 1229 (1972).
82. *Nieboer, E., McBryde, W. A. E.:* Can. J. Chem. *48*, 2565 (1970).

83. *Purdie, N., Farrow, M. M.:* Coord. Chem. Rev. *11*, 189 (1973); J. Solution Chem. *2*, 513 (1973).
84a. *Haas, Y., Navon, G.:* J. Phys. Chem. *76*, 1449 (1972).
84b. *Levine, B. A., Thornton, J. M., Williams, R. J. P.:* J. C. S. Chem. Commun. *1974*, 669.
85. *Graffeo, A. J., Bear, J. L.:* J. Inorg. Nucl. Chem. *30*, 1577 (1968).
86. *Silber, H. B., Farina, R. D., Swinehart, J. H.:* Inorg. Chem. *8*, 819 (1969).
87. *Geier, G., Karlen, U., Zelewsky, A. V.:* Helv. Chim. Acta *52*, 1967 (1969).
88. *Geier, G., Jørgensen, C. K.:* Chem. Phys. Letters *9*, 263 (1971).
89. *Kostromina, N. A., Tananaeva, N. N.:* Russ. J. Inorg. Chem. (English Transl.) *16*, 1256 (1971).
90. *Ternovaya, T. V., Kostromina, N. A.:* Russ. J. Inorg. Chem. (English Transl.) *18*, 190 (1973).
91. *Ots, H.:* Acta Chem. Scand. *27*, 2344 (1973).
92. *Ots, H.:* Acta Chem. Scand. *27*, 2351 (1973).
93. *Tananaeva, N. N., Kostromina, N. A.:* Russ. J. Inorg. Chem. (English Transl.) *14*, 631 (1969).
94. *Betts, R. H., Voss, R. H.:* Can. J. Chem. *51*, 538 (1973).
95. *Kostromina, N. A., Tananaeva, N. N.:* Russ. J. Inorg. Chem. (English Transl.) *16*, 673 (1969).
96. *Nieboer, E., Xavier, A. V., Williams, R. J. P., Rossotti, F. J. C., Dwek, R. A.:* unpublished results.
97. *Grenthe, I., Hessler, G., Ots, H.:* Acta Chem. Scand. *27*, 2543 (1973).
98. *Furie, B., Eastlake, A., Schechter, A. N., Anfinsen, C. B.:* J. Biol. Chem. *248*, 5821 (1973).
99. *Figgis, B. N., Lewis, J.:* In: Technique of inorganic chemistry, Vol. IV, p. 137 (*Jonassen, H. B., Weissberger, A.,* ed.). New York: Interscience 1965.
100. *Van Vleck, J. H.:* The Theory of electric and magnetic susceptibilities, Chapt. IX. Oxford: Oxford University Press 1932.
101. *Elliott, R. J.:* In: Magnetic properties of rare earth metals, Chapt. 1 (*Elliott, R. J.,* ed.). London: Plenium Press 1972.
102. *McMillan, J. A.:* Electron paramagnetism. New York: Reinhold 1968.
103. *Stewart, D. C., Kato, D.:* Anal. Chem. *30*, 164 (1958).
104. *Carnall, W. T., Fields, P. R.:* In: Lanthanide actinide Chemistry, Ch. 7 (*Fields, P. R., Moeller, T.,* ed.).
105. *Wybourne, B. G.:* Spectroscopic properties of rare earths. New York: Interscience 1965.
106. *Jørgensen, C. K., Judd, B. R.:* Mol. Phys. *8*, 281 (1964).
107. *Katzin, L. I.:* Inorg. Chem. *8*, 1649 (1969).
108. *Jørgensen, C. K.:* Mol. Phys. *5*, 271 (1962).
109. *Barnes, J. C.:* J. Chem. Soc. 3880 (1964).
110. *Nugent, L. J., Baybarz, R. D., Burnett, J. L.:* J. Phys. Chem. *77*, 1528 (1973).
111. *Ryan, J. L., Jørgensen, C. K.:* J. Phys. Chem. *70*, 2845 (1966).
112. *Ryan, J. L.:* Inorg. Chem. *8*, 2053 (1969).
113. *Filipescu, N., McAvoy, N.:* J. Inorg. Nucl. Chem. *28*, 253 (1966).
114. *Bhaumik, M. L., Telk, C. L.:* J. Opt. Soc. Am. *54*, 1211 (1964).
115. *Parker, C. A.:* Photoluminescence of solutions, p. 478. Amsterdam: Elsevier 1968.
116. *Gallagher, P. K.:* J. Chem. Phys. *41*, 3061 (1964).
117. *Kropp, J. L., Windsor, M. W.:* J. Chem. Phys. *39*, 2769 (1963).
118. *Charles, R. G., Riedel, E. P., Haverlack, P. G.:* J. Chem. Phys. *44*, 1356 (1966).
119. *Charles, R. G., Riedel, E. P.:* J. Inorg. Nucl. Chem. *28*, 527 (1966).
120. *Horrocks, Jr., Sipe, J. P., III, Sudnick, D.:* In: Nuclear magnetic resonance shift reagents, p. 53 (*Sievers, R. E.,* ed.). New York: Academic Press 1973.
121. *Eaton, D. R.:* In: Physical methods in advanced inorganic chemistry, p. 462 (*Hill, H. A. O., Day, P.,* eds.). London: Interscience 1968.
122. *Reuben, J., Fiat, D.:* J. C. S. Chem. Commun. *1967*, 729.
123. *Birnbaum, E. R., Moeller, T.:* J. Am. Chem. Soc. *91*, 7274 (1969).
124. *Bleaney, B.:* J. Magn. Reson. *8*, 91 (1972).
125. *Bleaney, B., Dobson, C. M., Levine, B. A., Martin, R. B., Williams, R. J. P., Xavier, A. V.:* J. C. S. Chem. Commun. *1972*, 791.

126. *Horrocks, W., DeW., Jr., Sipe, J. P., III:* Science *177*, 994 (1972).
127. *Barry, C. D., North, A. C. T., Glasel, J. A., Williams, R. J. P., Xavier, A. V.:* Nature *232*, 236 (1971).
128. *Podolski, M. L., Nieboer, E., Falter, H.:* unpublished results.
129. *Nieboer, E., East, D., Cohen, J. S., Schechter, A. N.:* unpublished results.
130. *Reuben, J., Leigh, J. S., Jr.:* J. Am. Chem. Soc. *94*, 2789 (1972).
131. *Horrocks, W., DeW.:* J. Am. Chem. Soc. *96*, 3022 (1974).
132. *Flora, W. P., Nieboer, E.:* unpublished results.
133. *Sternlicht, H.:* J. Chem. Phys. *42*, 2250 (1965).
134. *Dwek, R. A.:* Advan. Mol. Relaxation Precesses *4*, 1 (1972); NMR in Biochemistry. Oxford: Clarendon Press 1973.
135. *Dwek, R. A., Richards, R. E., Morallee, K. G., Nieboer, E., Williams, R. J. P., Xavier, A. V.:* European J. Biochem. *21*, 204 (1971).
136. *Bloembergen, N., Morgan, L. O.:* J. Chem. Phys. *34*, 842 (1961).
137. *Reuben, J., Reed, G. H., Cohn, M.:* J. Chem. Phys. *52*, 1617 (1970).
138. *Reuben, J.:* Biochemistry *10*, 2834 (1971).
139. *Reuben, J.:* J. Phys. Chem. *75*, 3164 (1971).
140. *Koenig, S. H.:* J. Chem. Phys. *56*, 3188 (1972).
141. *Bernheim, R. A., Brown, T. H., Gutowsky, H. S., Woessner, D. E.:* J. Chem. Phys. *30*, 950 (1959).
142. *Reuben, J.:* Naturwissenschaften *62*, 1975, in press.
143. *Mildvan, A. S., Cohn, M.:* Advan. Enzymol. *23*, 1 (1970).
144. *Reuben, J., Cohn, M.:* J. Biol. Chem. *245*, 6539 (1970).
145a. *Reed, G. H., Cohn, M.:* J. Biol. Chem. *245*, 662 (1970).
145b. *Case, G. D., Leigh, J. S., Jr.:* Proceedings of the Eleventh Rare Earth Conference, II, 706. Traverse City Mich.: U.S. Dept. of Commerce, Springfield, Va. October 1974; *Case, G. D.:* Biochim. Biophys. Acta., in press.
146. *Gerth, G., Kienle, P., Luchner, K.:* Phys. Letters *27 A*, 557 (1968).
147a. *Berkooz, O.:* J. Phys. Chem. Solids *30*, 1763 (1969).
147b. *Agarwala, U. C., Kumar, S., Rao, G. N., Medhi, O. K.:* Proceedings of the Eleventh Rare Earth Conference, II, 502. Traverse City Mich.: U.S. Dept. of Commerce, Springfield, Va. October 1974.
147c. *Spartalian, K., Oosterhuis, W. T.:* J. Chem. Phys. *59*, 617 (1973).
148. *Meites, L.:* Polarographic techniques. New York: Interscience 1965.
149. *Woyski, M. M., Harris, R. E.:* In: Treatise on analytical chemistry, Part II, Vol. 8, p. 1 (*Kolthoff, I. M., Elving, P. J.,* eds.). New York: Interscience 1963.
150. *Biedermann, G., Terjosin, G. S.:* Acta. Chem. Scand. *23*, 1896 (1969).
151. *Myasoedov, B. F., Sklyarenko, I. S., Kulyako, Y. M.:* Russ. J. Inorg. Chem. (English Transl.) *17*, 1541 (1972).
152. Standard methods of chemical analysis, Vol. III A (*Welcher, F. J.,* ed.). Princeton: D. Van Nostrand 1966.
153. *Dagnall, R. M., Smith, R., West, T. S.:* Analyst *92*, 358 (1967).
154. *Costanzo, D. A.:* Anal. Chem. *36*, 2042 (1964).
155. *Shults, W. D.:* Anal. Chem. *31*, 1095 (1959).
156. *Wilson, C. L., Wilson, D. W.:* Comprehensive analytical chemistry, Vol. I B, p. 356. Amsterdam: Elsevier 1960.
157. *Gallagher, P. K., Schrey, F., Prescott, B.:* Inorg. Chem. *9*, 215 (1970).
158. *Smolka, C. E., Birnbaum, E. R., Darnall, D. W.:* Biochemistry *10*, 4556 (1971).
159. *Choppin, G. R., Strazik, W. F.:* Inorg. Chem. *4*, 1250 (1965).
160. *Jones, A. D., Williams, D. R.:* J. Chem. Soc. (A) *1970*, 3138.
161. *Bertha, S. L., Choppin, G. R.:* Inorg. Chem. *8*, 613 (1969).
162. *Glasel, J. A.:* In: Progress in inorganic chemistry, Vol. 18, p. 383 (*Lippard, S. J.,* ed.). New York: Interscience 1973.
163. *Levitzki, A., Reuben, J.:* Biochemistry *12*, 41 (1973).
164. *Darnall, D. W., Birnbaum, E. R.:* Biochemistry *12*, 3489 (1973).
165. *Darnall, D. W., Birnbaum, E. R.:* J. Biol. Chem. *245*, 6484 (1970).

166a. *Darnall, D. W., Birnbaum, E. R., Gomez, J. E., Smolka, G. E.:* Proceedings of the Ninth Rare Earth Conference, I, p. 278. Blacksburg Va.: U.S. Dept. of Commerce, Springfield, Va. October 1971.

166b. *Izutsu, K. T., Felton, S. P., Siegal, I. A., Yoda, W. T., Chen, A. C. N.:* Biophys. Res. Commun. *49*, 1034 (1972).

167. *Kayne, M. S., Cohn, M.:* Biochem. Biophys. Res. Commun. *46*, 1285 (1972).

168. *Wedler, F. C., D'Aurora, V.:* Proceedings of the Tenth Rare Earth Conference, I, p. 137. Carefree, Arizona: U.S. Dept. of Commerce, Springfield, Va. May 1973.

169. *Fay, D. P., Litchinsky, D., Purdie, N.:* J. Chem. Phys. *73*, 544 (1969).

170. *Abrahamer, I., Marcus, Y.:* Inorg. Chem. *6*, 2103 (1967).

171. *Stroud, E. M., Kay, L. M., Dickerson, R. E.:* Cold Spring Harbor Symp. Quant. Biol. *36*, 125 (1971).

172. *Choppin, G. R., Friedman, H. G., Jr.:* Inorg. Chem. *5*, 1599 (1966).

173. *Darnall, D. W., Birnbaum, E. R., Sherry, A. D., Gomez, J. E.:* Proceedings of the Tenth Rare Earth Conference, I, p. 117. Carefree, Arizona: U.S. Dept. of Commerce, Springfield, Va. May 1973; Biochemistry *13*, 3745 (1974).

174. *Shoham, M., Kalb, A. J., Pecht, I.:* Biochemistry *12*, 1914 (1973).

175. *Sherry, A. D., Cottam, G. L.:* Proceedings of the Tenth Rare Earth Conference, II, 770. Carefree, Arizona: U.S. Dept. of Commerce, Springfield, Va. May 1973; Arch. Biochem. Biophys. *156*, 665 (1973). *Sherrey, A. D., Newman, A. D.:* Proceedings of the Eleventh Rare Earth Conference, I, 213. Traverse City Mich.: U.S. Dept. of Commerce, Springfield, Va. October 1974.

176. *Valentine, K. M., Cottam, G. L.:* Proceedings of the Tenth Rare Earth Conference, I, 127. Carefree, Arizona: U.S. Dept. of Commerce, Springfield, Va. May 1973; Arch. Biochem. Biophys. *158*, 346 (1973). *Cottam, G. L., Valentine, K. M., Thompson, B. C., Sherry, A. D.:* Biochemistry *13*, 3533 (1974).

177. *Sperow, J. W., Butler, L. G.:* Bioinorg. Chem. *2*, 87 (1973).

178. *Blaustein, M. P., Goldman, D. E.:* J. Gen. Physiol. *51*, 279 (1968).

179. *van Breemen, C., de Weer, P.:* Nature *226*, 760 (1970).

180. *van Breeman, C.:* Arch. Intern. Physiol. Biochem. *77*, 710 (1969).

181. *Weiss, G. B.:* Ann. Rev. Pharmacol. *14*, 343 (1974); *Weiss, G. B.:* Proceedings of the Tenth Rare Earth Conference, I, 149. Carefree, Arizona: U.S. Dept. of Commerce, Springfield, Va. May 1973; *Weiss, G. B., Goodman, F. R.:* Proceedings of the Eleventh Rare Earth Conference, II, 687, 716. Traverse City Mich.: U.S. Dept. of Commerce, Springfield, Va. October 1974.

182. *Twarog, B. M., Muneoka, Y.:* Cold Spring Harbor Symp. Quant. Biol. *37*, 489 (1972).

183. *Sperelakis, N., Valle, R., Orozco, C., Martinez-Palomo, A., Rubio, R.:* Am. J. Physiol. *225*, 793 (1973).

184. *Katzung, B. G., Reuter, H., Porzig, H.:* Experientia *29*, 1073 (1973).

185. *Chevallier, J., Butow, R. A.:* Biochemistry *10*, 2733 (1971).

186. *Anghileri, J.:* Intern. J. Clin. Pharmacol., Ther. Toxicol. *8*, 146 (1973).

187. *Lehninger, A. L., Carafoli, E.:* Arch. Biochem. Biophys. *143*, 506 (1971).

188. *Entman, M. L., Hansen, J. L., Cook, J. W., Jr.:* Biochem. Biophys. Res. Commun. *35*, 556 (1969).

189. *Peacocke, A. R., Williams, P. A.:* Nature *211*, 1140 (1966).

190. *Nagahashi, G., Thompson, W. W., Leonard, R. T.:* Science *183*, 670 (1974).

191. *Campbell, I. D., Dobson, C. M., Williams, R. J. P., Xavier, A. V.:* Proceedings of the Tenth Rare Earth Conference, II, P. 791. Carefree, Arizona: U.S. Dept. of Commerce, Springfield, Va. May 1973; Proceedings of the 5th Conference of "Magnetic Resonance in Biological Systems" New York Academy of Sciences, 1972; J. Magn. Resonance *11*, 172 (1973).

192. *Sieker, L. C., Adman, E., Jensen, L. H.:* Nature *235*, 40 (1972).

193. *Luk, C. K.:* Biochemistry *10*, 2838 (1971).

194a. *Epstein, M., Levitzki, A., Reuben, J.:* Biochemistry *13*, 1777 (1974).

194b. *Gafni, A., Steinberg, I. Z.:* Biochemistry *13*, 800 (1974).

194c. *Bunting, J. R., Cobo-Frenkel, A., Dowben, R. M.:* Proceedings of the Eleventh Rare Earth Conference, II, 672. Traverse City Mich.: U.S. Dept. of Commerce, Springfield, Va. October 1974.

194d. *Donato, H., Jr., Martin, R. B.:* Biochemistry *13*, 4575 (1974).

194e. *Cottam, G. L., Sherry, A. D., Valentine, K. M.:* Proceedings of the Eleventh Rare Earth Conference, I, 204. Traverse City Mich.: U.S. Dept. of Commerce, Springfield, Va. October 1974.

194f. *Dwek, R. A., Knott, J. C. A., Marsh, D., McLaughlin, A. C., Press, E. M., Price, N. C., White, A. I.:* Proceedings of the Eleventh Rare Earth Conference, I, 184; J. European Biochem, in press.

194g. *Kayne, M. S., Cohn, M.:* Biochemistry *13*, 4159 (1974).

194h. *Formosa, C.:* Biochem. Biophys. Res. Commun. *53*, 1084 (1973).

194i. *Eisinger, J., Lamola, A. A.:* Biochim. Biophys. Acta *240*, 299, 313 (1971).

194j. *Mikkelsen, R. B., Wallach, D. F. H.:* Proceedings of the Eleventh Rare Earth Conference, II, 698. Traverse City Mich.: U.S. Dept. of Commerce, Springfield, Va. October 1974.

194k. *Mahler, H. R., Cordes, E. H.:* Biological chemistry, p. 128. New York: Harper and Row 1966.

194l. *Choppin, G. R., Fellows, R. L.:* J. Coord. Chem. *3*, 209 (1974).

194m. *Henrie, D. E., Henrie, B. K.:* J. Inorg. Nucl. Chem. *36*, 2124 (1974).

194n. *Prados, R., Stadtherr, L. G., Donato, Jr., H., Martin, R. B.:* J. Inorg. Nucl. Chem. *36*, 689 (1974).

194o. *Sipe, J. P., Martin, R. B.:* J. Inorg. Nucl. Chem. *36*, 2122 (1974).

194p. *Abbott, F., Darnall, D. W., Birnbaum, E. R.:* Reported at the Eleventh Rare Earth Conference, Traverse City Mich. October 1974. See also Conference Proceedings. I, 194: U.S. Dept. of Commerce, Springfield, Va.

194q. *Carver, J. P., Barber, B. H., Quirt, A.:* Reported at the VIth International Conference on Magnetic Resonance in Biological Systems. Abstract C37: Kandersteg, Switzerland, September, 1974.

195. *Scatchard, G.:* Ann. N. Y. Acad. Sci. *51*, 660 (1949).

196. *Nakazawa, E., Shionoy, S.:* J. Chem. Phys. *47*, 3211 (1967).

197. *Latt, S. A., Cheung, H. T., Blout, E. R.:* J. Am. Chem. Soc. *87*, 995 (1965).

198. *Birnbaum, E. R., Gomez, J. E., Darnall, D. W.:* J. Am. Chem. Soc. *92*, 5287 (1970).

199. *Birnbaum, E. R., Yoshida, C., Gomez, J. E., Darnall, D. W.:* Proceedings of the Ninth Rare Earth Conference, I, p. 264. Blacksburg Va.: U.S. Dept. of Commerce, Springfield, Va. October 1971.

200. *Birnbaum, E. R., Sherry, A. D., Darnall, D. W.:* Proceedings of the Tenth Rare Earth Conference, II, p. 753. Carefree Arizona: U.S. Dept. of Commerce, Springfield, Va. May 1973; Bioinorg. Chem. *3*, 15 (1973).

201. *Karraker, D. G.:* Inorg. Chem. *7*, 473 (1968).

202. *Sherry, A. D., Yoshida, D., Birnbaum, E. R., Darnall, D. W.:* J. Am. Chem. Soc. *95*, 3011 (1973).

203. *Sherry, A. D., Birnbaum, E. R., Darnall, D. W.:* J. Biol. Chem. *247*, 3489 (1972).

204. *Farrar, C. T., Becker, E. D.:* Pulse and fourier transform NMR. New York: Academic Press 1971.

205. *Dwek, R. A., Ferguson, S. J., Radda, G. K., Williams, R. J. P., Xavier, A. V.:* Proceedings of the Tenth Rare Earth Conference, I, p. 111. Carefree, Arizona: U.S. Dept. of Commerce, Springfield, Va. May 1973.

206. *Nowak, T., Mildvan, A. S., Kenyon, G. L.:* Biochemistry *12*, 1690 (1973).

207. *Nieboer, E., East, D., Cohen, J. S., Furie, B., Schechter, A. N.:* Proceedings of the Tenth Rare Earth Conference, II, p. 763. Carefree, Arizona: U.S. Dept. of Commerce, Springfield, Va. May 1973.

208. *Furie, B., Griffen, J. H., Feldman, R., Sokoloski, E. A., Schechter, A. N.:* Proc. Natl. Acad. Sci. U.S. *71*, 2833 (1974).

209. *Nieboer, E., Flora, W. P., Podolski, M., Falter, H.:* Proceedings of the Tenth Rare Earth Conference, I, p. 388. Carefree, Arizona: U.S. Dept. of Commerce, Springfield, Va. May 1973.

210. *Tanner, S. P., Choppin, G. R.:* Inorg. Chem. *7*, 2046 (1962).

211. *Butler, J. N.:* Ionic equilibrium, p. 321. Reading, Mass.: Addison-Wesley 1964.
212. *Bystrov, V. F., Dubrovina, N. I., Barsukov, L. I., Bergelsen, L. D.:* Chem. Phys. Lipids *6*, 343 (1971).
213. *Kostelink, R. J., Castellano, S. M.:* J. Magn. Resonance *7*, 219 (1972).
214. *Huang, C.:* Biochemistry *8*, 344 (1969).
215. *Huang, C., Sipe, J. P., Chow, S. T., Martin, R. B.:* Proc. Natl. Acad. Sci. US *71*, 359 (1974).
216. *Michaelson, D. M., Horwitz, A. F., Klein, M. P.:* Biochemistry *12*, 2637 (1973).
217. *Assmann, G., Sokoloski, E. A., Brewer, H. B., Jr.:* Proc. Natl. Acad. Sci. U.S. *71*, 549 (1974).
218. *Williams, R. J. P.:* In: Proc. 4th Int. Congr. Pharmacol. *5*, 227 (*Eigenmann, R.,* ed.). Basel, Switzerland: Schwabe 1969.
219. *Andrews, S. B., Faller, J. W., Gilliam, J. M., Barrnett, R. J.:* Proc. Natl. Acad. Sci. U.S. *70*, 1814 (1973).
220. *Barry, C. D., Glasel, J. A., Williams, R. J. P., Xavier, A. V.:* J. Mol. Biol. *84*, 471 (1974).
221. *Barry, C. D., Martin, D. R., Williams, R. J. P., Xavier, A. V.:* J. Mol. Biol. *84*, 491 (1974).
222. *Dobson, C. M., Williams, R. J. P., Xavier, A. V.:* J. C. S. Dalton *1973*, 2662.
223. *Dobson, C. M., Williams, R. J. P., Xavier, A. V.:* J. C. S. Dalton *1974*, 1762. 1765.
224. *Lavallee, D. K., Zeltman, A. H.:* Proceedings of the Eleventh Rare Earth Conference, I, 218. Traverse City Mich.: U.S. Dept. of Commerce, Springfield, Va. October 1974.
225. *Barry, C. D., Dobson, C. M., Sweigart, S. A., Ford, L. E., Williams, R. J. P.:* In: Nuclear magnetic resonance shift reagents, p. 173 (*Sievers, R. E.,* ed.). New York: Academic Press 1973.
226. *Harrison, M. R., Moulds, B. E., Rossotti, F. J. C.:* Trans. Royal Inst. Technol. (Stockholm) *268*, 263 (1972).
227. *Levine, B. A., Williams, R. J. P.:* Proc. Roy. Soc. (London), 1975, in press.
228. *Jones, R., Dwek, R. A., Forsén, S.:* European J. Biochem. *47*, 271 (1974).
229. *Dobson, C. M., Campbell, I. D., Williams, R. J. P.:* Proc. Roy. Soc. (London), 1975, in press.
230. *Tanswell, P., Westhead, E. W., Williams, R. J. P.:* Biochem. Soc. Trans. *1*, 79 (1974).
231. *Tanswell, P., Westhead, E. W., Williams, R. J. P.:* Febs Letters *48*, 60 (1974).
232. *Dwek, R. A., Levy, H. R., Radda, G. K., Seeley, P. J.:* Biochem. Biophys. Acta *377*, 26 (1975).
233. *Dobson, C. M., Moore, G. R., Williams, R. J. P.:* Febs. Letters *51*, 60 (1975).
234. *Anteunis, M., Gelan, J.:* J. Am. Chem. Soc. *95*, 6502 (1973)
235. *Bradbury, J. H., Brown, L. R., Crompton, M. W., Warren, B.:* Pure and Applied Chemistry *40*, 83 (1974); Analyt. Biochem. *62*, 310 (1974).

Received June 14, 1974 (April 22, 1975).

Partly Filled Shells Constituting Anti-bonding Orbitals with Higher Ionization Energy than their Bonding Counterparts

C. K. Jørgensen

Département de Chimie minérale et analytique, Université de Genève, CH-1211 Geneva 4, Switzerland

Table of Contents

I. One-electron Substitutions in the Lowest Configuration

A. Monatomic Entities

It is an empirical fact that the lowest 20 to 400 J-levels of gaseous atoms or of positive ions M^{+z} can be *classified* (*1*) as belonging to *electron configurations*. Certain atoms have all their levels described by the nl-value of an external electron outside closed-shells. For instance, the sodium atom has the groundstate $1s^2 2s^2 2p^6 3s$ usually abbreviated [Ne] $3s$. The first excited configuration [Ne] $3p$ contains two closely adjacent J-levels (producing the two spectral lines in the yellow, $3p \to 3s$ in emission and $3s \to 3p$ in absorption) and the subsequent configurations [Ne] $4s$, [Ne] $3d$, [Ne] $5s$, [Ne] $4d$, [Ne] $4f$, [Ne] $6s$, ... were first described by Rydberg as having the energy $- (z+1)^2/(n-d)^2$ times the Rydberg constant 109737 cm$^{-1} = 13.60$ eV, where z is the ionic charge (zero for neutral atoms), n is the principal quantum number, and d is the *Rydberg defect* which normally is *almost* constant as a function of n for a given value of the orbital angular quantum number l. Thus, $d = 1.37$ for [Ne] ns, 0.88 for [Ne] np, 0.01 for [Ne] nd and 0.00 for [Ne] nf where the energies are given relative to the ionization limit of the system containing one external electron, the zero-point here representing the groundstate of Na$^+$. It is taken as an indication of no penetration of the external electron (Leuchtelektron) in the closed shells of the core (Atomrumpf) when d is negligible, as it is for f-electrons until xenon ($Z = 54$) and for g-electrons in the known part of the Periodic Table (Z below 118). The situation is somewhat different in the [Xe] nl systems, for which the Rydberg defect d is (*2*):

	[Xe] $6s$	[Xe] $6p$	[Xe] $5d$	[Xe] $4f$	
Cs	4.13	3.63	2.45	0.02	
Ba$^+$	3.67	3.28	2.59	0.31	(1)
La^{+2}	3.36	3.00	2.47	1.40	
Ce^{+3}	3.11	2.80	2.32	1.56	
Pr^{+4}	2.89	2.59	2.19	1.56	

Like in the isoelectronic series (*3*) K, Ca$^+$, Sc^{+2}, ... where the [Ar] $4s$ is below [Ar] $3d$ in the potassium atom and in Ca$^+$ but $3d$ becomes progressively more stable from Sc^{+2} onward, the [Xe] $5d$ supplies the groundstate of La^{+2} but looses the competition with [Xe] $4f$ from Ce^{+3} on. However, contrary to the cases of the alkaline metals, the $5f$ electrons do not show the same d as $4f$ but rather 1.03 both in the beginning of the lanthanides, and for elements distributed between Lu^{+3} ($Z = 71$) and the mercury atom ($Z = 80$). This deviation from the Rydberg formula is quite characteristic for the transition groups.

Other monatomic entities are more complicated. *Paschen* showed in 1919 that the many J-levels of the neon atom can be described as the excitation of a $2p$ electron to an empty nl-shell. Thus, the next-lowest configuration $1s^2 2s^2 2p^5 3s$ contains 4 J-levels and the following $1s^2 2s^2 2p^5 3p$ ten J-levels. The 30 allowed transitions between these two configurations give the characteristic spectral lines in the red of neon. Higher configurations such as $1s^2 2s^2 2p^5 3d$ or $1s^2 2s^2 2p^5 4f$ can also be recognized. As pointed out by *Humphreys* and *Meggers* (*4*) in the case of krypton and xenon, and by *Ebbe Rasmussen* (5) in the case of radon, it is characteristic that the two J-values ($3/2$ and $1/2$) possible for the core terminating np^5 produce two sets of excited levels when an external electron is added (such as $5p^5\,5d$ or $5p^5\,4f$ in the case of xenon) with an energy difference almost identical with energy difference between the two J-values, 5371 cm^{-1} in Kr$^+$, 10537 cm^{-1} in Xe$^+$ and 30895 cm^{-1} in Rn$^+$, constituting two different ionization limits for the series spectra varying $n = 5, 6, 7, \ldots$ and the energy of the external electron follows the *Rydberg* formula to a good approximation.

Systems with two external electrons can show series converging to ionization limits corresponding to excited levels of the ionized system with one remaining external electron. *Henry Norris Russell* and *Saunders* discovered in 1925 that certain levels of the calcium atom belong to configurations [Ar] $3d\,4d$ or [Ar] $3d\,5p$ but have *higher* energy than the (lowest) ionization energy producing the ground-state [Ar] $4s$ of Ca$^+$ whereas the two J-levels of [Ar] $3d$ have 13650 and 13711 cm^{-1} higher energy. Hence, [Ar] $3d$ provides higher ionization limits in the same sense as the excited $^2P_{1/2}$ level of the monopositive, heavier noble gases. This existence of *auto-ionizing levels* in the *continuum* above the first ionization energy has a certain connection with the photo-electron spectra discussed below. Most configurations of the copper atoms are [Ar] $3d^{10}(nl)^1$ or [Ar] $3d^9\,4s(nl)^1$ also showing adjacent ionization limits being J-levels of Cu$^+$.

The typical transition group atom has many more low-lying levels producing "multiplet spectra" with numerous spectral lines, and the individual J-levels can be connected in terms with definite values of S and L in the case of approximate validity of Russell-Saunders coupling. Thus, the neutral iron atom (producing the majority of the stronger Fraunhofer lines seen in absorption in the Solar spectrum) has the groundstate and 22 other J-levels (among the 34 predicted in Hund's vector-coupling model) up to 29799 cm^{-1} belonging to [Ar] $3d^6\,4s^2$, but already at 6928 cm^{-1} starts the configuration [Ar] $3d^7\,4s$ and at 32874 cm^{-1} [Ar] $3d^8$. The configurations with *odd parity* (having the sum of l-values odd) are [Ar] $3d^6\,4s\,4p$ (containing the lowest level at 19351 cm^{-1}), [Ar] $3d^7\,4p$ and many others.

When comparing intensities of the spectral lines, the atomic spectroscopists soon realized that the selection rules such as $J \to (J-1), J, (J+1)$ and the most important (that transitions only can be allowed between levels of opposite parity) are incomplete insofar the transitions usually have negligible intensity (and are difficult to detect) if the configurations do not differ in exactly one electron, and further on, that this electron changes l by one unit. This rule (which is an approximation the same way as the selection rule for *spin-allowed transitions* that the total spin quantum number S is invariant) supported the general feeling that the total wave-functions Ψ of each level (with rare exceptions) must be well represented by well-defined electron configurations. However, this is definitely not the case (6) in the strict sense that Ψ is a linear combination (with coefficients prescribed (7) by the combination of S and L in Russell-Saunders coupling) of anti-symmetrized Slater determinants. There are good reasons to believe that the overlap integral of the groundstate Ψ of the argon atom with the unique Slater determinant corresponding to $1s^2\,2s^2\,2p^6\,3s^2\,3p^6$ is below 0.5. The origin of this, somewhat paradoxial, situation is that Ψ of 18 electrons is extended in a 54-

dimensional space. Though the electron density is nicely described in our three-dimensional space by the Hartree-Fock solution, the interelectronic repulsion (integrated over 6 spatial variables at a time) is overestimated, and must be corrected by a kind of dielectric screening (6). Anyhow, the rapid decrease in the number of atomic spectroscopists since nuclear transmutations became fashionable in 1931 has hidden the opposite side of this dilemma, that the classificatory success of electron configurations is remarkable. A rather extreme case is Pr^{+2} (8, 9) where 38 of the 41 levels of the lowest configuration [Xe] $4f^3$ have been found between 0 and 39941 cm^{-1} and 101 of the 107 levels of the next configuration [Xe] $4f^2\,5d$ between 12847 and 48745 cm^{-1} above the groundstate. Another example is *Bryant* (10) finding all 20 levels of [Xe] $4f^{13}\,5d$, the 4 levels of [Xe] $4f^{13}\,6s$ and all 12 levels of [Xe] $4f^{13}\,6p$ of Yb^{+2} (having the closed-shell groundstate [Xe] $4f^{14}$). The same complete configurations have been found by *Sugar* and *Kaufman* (11) in the isoelectronic Lu^{+3} and (127) in Hf^{+4}.

The neutral lanthanide atoms have extremely dense distributions of spectral lines, and it is difficult to perform a significant analysis without a high percentage of accidental coincidences. From this point of view, gaseous M^{+2} and M^{+3} (12, 13) are somewhat simpler. For instance, they have very characteristic groups of lines in the ultra-violet due to emission from [Xe] $4f^q\,6p$ to [Xe] $4f^q\,6s$. Though each of these configurations are calculated to be some 50000 to 100000 cm^{-1} wide, the intense transitions go between levels with almost the same parentage (S, L, J) of the $4f^q$ part of the wave-function in the two cases. This selectivity reminds one about the conditions in xenon and radon (5) where it is different to find combinations between the two systems belonging to $np^5(J = 3/2$ and $1/2)$ parentage.

There is an aspect of atomic spectroscopy, which is frequently neglected by chemists believing in intrinsically additive one-electron energies. To the first approximation, the barycentre (energy weighted by the number of states $(2J + 1)$ in each J-level) of the configuration containing two partly filled shells $a^m\,b^n$ can be written (2, 3, 14)

$$- m\,I_\varepsilon(a) - n\,I_\varepsilon(b) + \frac{m(m-1)}{2}\,A_*(a, a) + mn\,A_*(a, b) + \frac{n(n-1)}{2}\,A_*(b, b) \quad (2)$$

where the two I_ε values are the ionization energy of the system containing only one electron besides the closed shell (the zero-point of energy of eq. (2) corresponds to these closed shells alone) and the average parameters A_* of *interelectronic repulsion* are defined (3) as a linear combination of *Slater-Condon-Shortley* (7) or *Racah* (15) parameters for a given l value. If $2\,A_*(a, b) = A_*(a, a) + A_*(b, b)$, Eq. (2) would predict a linear variation of the total energy as a function of the occupation numbers m and n for constant ionic charge, and hence constant $(m + n)$. However, in actual practice, one of the shells have a considerably larger average radius than the other shell, let it be b. Then, $A_*(a, a) \gg A_*(a, b) \sim A_*(b, b)$ with the result that Eq. (2) indicates a *parabolic* variation with the occupation numbers m and n. This is in agreement with experience (14, 16) for [Ar] $3d^m\,4s^n$ where the barycentre of [Ar] $3d\,4s$ of Sc^+ is situated at 0.09 eV above the groundstate (one of the two J-levels of this configuration) at about equal distance *below* [Ar] $3d^2$ having the barycentre at 1.24 eV (1 eV $= 8066$ cm^{-1}) and the unique state [Ar] $4s^2$ at 1.45 eV. When combined with $I_\varepsilon(3d) = 24.75$ and $I_\varepsilon(4s) = 21.60$ eV known from Sc^{+2}, these barycenters *(14, 16)* suggest $A_*(3d, 3d) = 13.56$, $A_*(3d, 4s) = 8.38$ and $A_*(4s, 4s) = 6.60$ eV assuming Eq. (2). Another example of minimum energy for $n = 1$ is the gaseous nickel atom, where the barycentre of [Ar] $3d^8\,4s^2$ occurs at 1.29 eV, though the groundstate $^3F_4$ belongs to this configuration, whereas the lowest barycentre is [Ar] $3d^9\,4s$ at 0.17 eV, and the unique state of [Ar] $3d^{10}$ at 1.83 eV above the groundstate. This situation differs in the palladium atom, where the groundstate [Kr] $4d^{10}$ is 1.08 eV below the barycentre of [Kr] $4d^9\,5s$, and [Kr] $4d^8\,5s^2$ is ~ 4 eV above the closed-shell groundstate.

All the neutral $3d$ group atoms (with exception of [Ar] $3d^5\,4s$ in chromium and [Ar] $3d^{10}\,4s$ in copper) contain two $4s$ electrons in the groundstate, as if the $4s$ electrons are more stable than the $3d$ electrons. However, this argument based on occupation numbers of the shells is not satisfactory to the chemist, because the ionization energy of $4s$ is distinctly below that of $3d$, and all the gaseous M^{+2}, M^{+3}, M^{+4}, ... of the $3d$ group have the configuration [Ar] $3d^q$ well below all others. In sofar the great majority of transition group compounds *(3, 17—19)* have a preponderant electron configuration with *one* recognizable, partly filled d or f shell, the corresponding *Aufbau principle* is important to chemists:

$1s \ll$	[He]:	H($-$I), He, Li(I), Be(II), B(III), C(IV), N(V)
$2s < 2p \ll$	[Ne]:	C($-$IV), N($-$III), O($-$II), F($-$I), Ne, Na(I), Mg(II), Al(III), Si(IV), P(V), S(VI), Cl(VII)
$3s < 3p \ll$	[Ar]:	P($-$III), S($-$II), Cl($-$I), Ar, K(I), Ca(II), Sc(III), Ti(IV), V(V), Cr(VI), Mn(VII)
$3d <$	[28]:	Fe($-$II), Co($-$I), Ni(0), Cu(I), Zn(II), Ga(III), Ge(IV), As(V), Se(VI), Br(VII)
$4s < 4p \ll$	[Kr]:	As($-$III), Se($-$II), Br($-$I), Kr, Rb(I), Sr(II), Y(III), Zr(IV), Nb(V), Mo(VI), Tc(VII), Ru(VIII)
$4d <$	[46]:	Ru($-$II), Rh($-$I), Pd(0), Ag(I), Cd(II), In(III), Sn(IV), Sb(V), Te(VI), I(VII), Xe(VIII)
$5s <$	[48]:	In(I), Sn(II), Sb(III), Te(IV), I(V), Xe(VI)
$5p \ll$	[Xe]:	Sb($-$III), Te($-$II), I($-$I), Xe, Cs(I), Ba(II), La(III), Ce(IV)
$4f <$	[68]:	Yb(II), Lu(III), Hf(IV), Ta(V), W(VI), Re(VII), Os(VIII)
$5d <$	[78]:	Os($-$II), Ir($-$I), Pt(0), Au(I), Hg(II), Tl(III), Pb(IV), Bi(V)
$6s <$	[80]:	Au($-$I), Hg, Tl(I), Pb(II), Bi(III), Po(IV)
$6p \ll$	[Rn]:	Bi($-$III), Po($-$II), At($-$I), Rn, Fr(I), Ra(II), Ac(III), Th(IV), Pa(V), U(VI), Np(VII)
$5f <$	[100]:	No(II), Lr(III), 104(IV), 105(V)

$$(3)$$

The closed shells corresponding to noble gases are indicated by double inequality signs, but other closed shells of interest to chemists are marked with the number of electrons in square brackets. It is seen that most of the isoelectronic series have from 7 to 12 members, though it must be realized that many of the oxidation states given in Eq. (3) are far from being the most frequent for a given element. As far goes gaseous M^{+2}, M^{+3}, ... this Aufbau principle has five exceptions: La^{+2} [Xe] $5d$; Gd^{+2} [Xe] $4f^7\,5d$; Lu^{+2} [Xe] $4f^{14}\,6s$; Ac^{+2} [Rn] $7s$ and

Th^{+2} [Rn] $5f\,6d$ (128). In all cases, the expected configurations ([Xe] $4f$; [Xe] $4f^8$; [Xe] $4f^{14}\,5d$; [Rn] $5f$ and [Rn] $5f^2$) have been detected as relatively low-lying levels. An incomplete list of these exceptions was given in a previous review (20). Another correction is the mis-spelling of the name of *Stoner*, who was the first, in 1924 (21), to suggest that each nl-shell contains ($4l+2$) electrons.

For neutral atoms, a much more approximative Aufbau principle is

$$1s \ll 2s < 2p \ll 3s < 3p \ll 4s < 3d < 4p \ll 5s < 4d < 5p \ll$$
$$6s < 4f < 5d < 6p \ll 7s < 5f < 6d < \ldots \tag{4}$$

allowing twenty exceptions (Cr, Cu, Nb, Mo, Tc, Ru, Rh, Pd, Ag, La, Ce, Gd, Pt, Au, Ac, Th, Pa, U, Np and Cm) among which only palladium and thorium differ in two electrons from the consecutive order in Eq. (4). However, the question of the exact configuration providing the groundstate of the gaseous atom may be less important than believed by many chemists. Thus, the groundstate of the terbium atom (22) belonging to [Xe] $4f^8$ is only 286 cm^{-1} below the lowest level belonging to [Xe] $4f^7\,5d$, which is negligible in view of the fact that these two configurations both have their levels dispersed over more than 80000 cm^{-1}.

B. 4f Group Compounds

Whereas configurations such as [Xe] $4f^q\,6s^2$, [Xe] $4f^{q-1}\,5d\,6s$ and [Xe] $4f^{q-1}\,5d^2$ strongly overlap in neutral lanthanide atoms (23—28) and M$^+$ usually has the groundstate belonging to [Xe] $4f^q\,6s$ followed by other configurations such as [Xe] $4f^q\,5d$ and [Xe] $4f^{q-1}\,5d\,6s$ (the configuration [Xe] $4f^{q+1}$ has not usually been detected; according to Eq. (2), it is expected to have high energy) there is no serious problem with M^{+2} (excepting Gd^{+2}) and M^{+3} all having [Xe] $4f^q$ groundstates. This is also true to a remarkable accuracy for all known M(II) and M(III) compounds, unless they are metallic, and also excepting the somewhat enigmatic behaviour (29, 30) of La(II), Ce(II), Gd(II) and Tb(II) in fluorite(CaF$_2$) crystals.

In monatomic entities, the *system difference* is the energy difference (20, 23— 28) between the lowest level of [Xe] $4f^{q-1}\,5d\,6s^2$ and of [Xe] $4f^q\,6s^2$ in neutral atoms, between [Xe] $4f^{q-1}\,5d\,6s$ and [Xe] $4f^q\,6s$ in M$^+$, and between the lowest level of [Xe] $4f^{q-1}\,5d$ and the lowest level of [Xe] $4f^q$ in gaseous M^{+2} and M^{+3} and in M(II) and M(III) compounds. One might have expected that each of these configurations would show a very complicated distribution of J-levels depending on a large number of parameters (7, 15) of interelectronic repulsion, but it turns out that the system difference is essentially determined by the energy difference between the lowest level of the ionized system $4f^{q-1}$ and of $4f^q$, from which is subtracted a large contribution varying very smoothly as a function of the atomic number Z. Correspondingly, the system difference (see Fig. 1) varies in a characteristic way as a function of q, showing two zigzag-curves beginning at $q=1$ and 8 and having their highest points at $q=7$ and 14, and having almost the same value for q and $(7+q)$ which is an almost unique property of the $4f$ group, though *Catalán*, *Röhrlich* and *Shenstone* (31) previously had made similar observations of [Ar] $3d^{q-1}\,4s$ and [Ar] $3d^q$ in monatomic 3d group entities. The system difference increases $M^0 \lesssim M(II) < M^+ < M^{+2} \ll M(III) < M^{+3} \ll M^{+4}$ for a given value of q, showing a vertical parallel translation on Fig. 1.

The two zigzag curves can be explained by the *refined spin-pairing energy theory (20, 32, 33)* which was originally intended to explain the electron transfer spectra of M(III) complexes *(34)*. Relative to the barycentre of all the states of [Xe] $4f^q$ described by Eq. (2) the barycentre of each manifold of such states having a definite value of the total spin quantum number S has the energy *(3, 6, 19)*:

$$D \frac{3q(4l + 2 - q)}{16l + 4} - S(S + 1) D \qquad (5)$$

where the spin-pairing energy parameter D is a definite linear combination of the parameter F^2, F^4 and F^6 of interelectronic repulsion, or $\frac{9}{8} E^1$ according to *Racah (15)* for f electrons. The numerical value of D is close to 6500 cm^{-1} or 0.8 eV. The complicated fraction to the left in Eq. (5) is the average value of $<S(S + 1)>$ for the whole configuration l^q. The difference between the barycentre of states with $(S - 1)$ and with S is $2 D S$. Considering (Eq. 5) alone, the system differences on Fig. 1 should be represented by two straight lines with the slope $(8 D/13)$ per unit of q and showing a vertical shift $8 D$ between $q = 7$ and 8. However, some f^q have more than one term possessing the maximum value of S, with the result that the lowest terms (^3H of f^2 and f^{12} and ^6H of f^5 and f^9) are situated $9 E^3$ below the barycentre given by Eq. (5) and the terms with

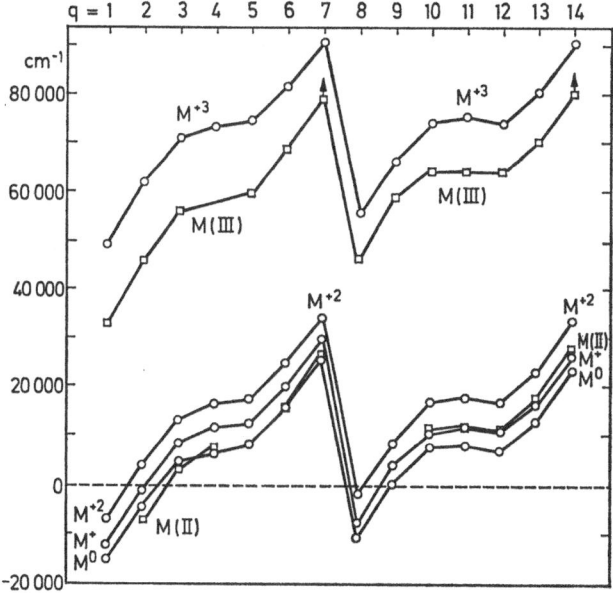

Fig. 1. The energy differences between the lowest J-level of two configurations differing in one $5d$ electron becoming a $4f$ electron, so the lower state (when the system difference is positive) contains $4f^q$. In the nine cases of negative system difference, the lower state contains $4f^{q-1} 5d$. The squares represent ions in fluorite.

$L = 5$ (^4I of f^3 and f^{11} and ^5I of f^4 and f^{10}) are stabilized 21 E^3. The *Racah* parameter E^3 is about a-tenth of E^1. Further on, the first-order relativistic effect expressed by the *Russell-Saunders* asymptotic value *(14)* of spin-orbit coupling stabilizes the lowest J-level of the lowest (S, L) term to the extent $(\frac{1}{2} L + \frac{1}{2}) \zeta_{4f}$ in the first half of the $4f$ shell ($q < 7$) and $\frac{1}{2} L \zeta_{4f}$ for q above 7. The *Landé* parameter ζ_{4f} increases smoothly from 0.08 eV for Ce(III) to 0.36 eV for Yb(III). The combination of these two effects modify the curves on Fig. 1 from two parallel line segments to the characteristic zigzag with almost horizontal plateaux at $q = 3$, 4, 5 and $q = 10$, 11, 12.

Both in gaseous monatomic entities and in M(II) and M(III) compounds, the general stabilization of the $4f$ shell going from one element to the next (corresponding to the difference (E—A) between stronger nuclear attraction $- (Z + 1)/r$ and increased interelectronic repulsion) produce comparable system differences for $q = 7$ and 14 because $7(E-A)$ varying between 2.5 and 2.7 eV has the same order of magnitude as $(48 D/13) - \frac{3}{2} \zeta_{4f}$. This is not exactly the case for the *ionization energy* I_3 of gaseous M^{+2} according to *Sugar* and *Reader* *(35)* which is 24.70 eV for Eu^{+2} and 25.03 eV for Yb^{+2} suggesting $7(E-A)$ close to 2.8 eV, nor for I_4 of gaseous M^{+3} being 44.01 eV for Gd^{+3} and 45.19 eV for Lu^{+3} where $7(E-A)$ is 3.6 eV unless D is considerably larger. The variation of I_4 from 39.79 eV for Tb^{+3} ($q = 8$) to $q = 14$ consisting essentially of $6(E-A) + (48/13) D + \frac{3}{2} \zeta_{4f}$ actually can be combined with the former difference to give $(E-A) = 0.47$ eV and $D = 0.72$ eV, rather showing the opposite trend. Anyhow, it is fascinating how the variation with q predicted by the refined spin-pairing energy theory applies to so different quantities as I_3 and I_4 of the gaseous ions and the [Xe] $4f^q \rightarrow$ [Xe] $4f^{q-1} 5d$ excitation energies of M(II) *(29)* and M(III) *(36)* substituted in CaF_2. Whereas fluorite is transparent down to 1100 Å or 11 eV, water is only transparent to 1600 Å or 8 eV with the result that the $4f \rightarrow 5d$ transitions at slightly higher wave-numbers than in CaF_2 have only been observed *(37, 38)* in Ce(III), Pr(III) and Tb(III) aqua ions.

Many spectroscopists tend to consider only the internal transitions in the partly filled $4f$ shell and inter-shell transitions such as $4f \rightarrow 5d$, but disregard the possibility of inter-atomic transitions in M(III) and M(IV) compounds. The *electron transfer spectra* are due to excited states *(6, 39)* where an electron has been transferred from one or more reducing ligands X to the empty or partly filled shell of the central atom M. It is possible to define *optical electronegativities* x_{opt} from the wave-number $\nu_{e.t.}$ of the first *Laporte*-allowed electron transfer band. For our purpose, it is useful to consider the uncorrected optical electronegativity x_{uncorr} (M) without corrections for sub-shell energy differences (in d groups) and from the spin-pairing energy theory, in which case

$$\nu_{e.t.} = [x_{opt} (X) - x_{uncorr} (M)] \cdot 30000 \ cm^{-1} \tag{6}$$

where $x_{opt} = 3.9$ for fluoride, 3.0 for chloride, 2.8 for bromide and 2.5 for iodide, corresponding to the *Pauling* electronegativities. In the $4f$ group, $\nu_{e.t.}$ increases for a given ligand Eu(III) $<$ Yb(III) $<$ Sm(III) $<$ Tm(III) $< \ldots$ corresponding to Fig. 1 turned upside down, and the spin-pairing theory was introduced *(34)* for explaining the electron transfer spectra of bromide complexes in almost anhydrous ethanol. The hexabromides MBr_6^{-3} studied in acetonitrile (40) has lower $\nu_{e.t.}$ and the lowest wave-numbers are found (41) in the chemically rather unstable hexa-iodides MI_6^{-3}.

Another type of system difference are the *standard oxidation potentials* of M(II) aqua ions E^0 which can also be described (33) by the refined spin-pairing energy theory, being the least negative for Eu(II). A corresponding set of highly positive E^0 have been calculated (33) for M(III) aqua ions, the lowest value being for Ce(III). Since the $-\Delta G$ and $-\Delta H$ for the hydration of a gaseous proton (42) is close to 11.3 eV, it is possible (3, 43) to define a *chemical ionization energy*

$$I_{chem} = E^0 + 4{,}5 \text{ eV} \tag{7}$$

where the constant 4.5 eV (corresponding to the ionization energy of the standard hydrogen electrode to expel an electron in *vacuo*) is known within a-tenth eV. The *hydration energy difference* (43) $I_z - I_{chem}$ for the one-electron oxidation of the $(z-1)$ aqua ion turns out to be $(2z-1)\varkappa$ where the parameter $\varkappa$ is 5.3 eV in the 3d group and 4.3 eV in the 4f group. Thus, I_{chem} for Eu(II) aqua ions is calculated to $24{,}7 - 5 \cdot 4.3 = 3.2$ eV to be compared with the measured $I_{chem} = 4.1$ eV, whereas I_{chem} for Ce(III) aqua ions should be $36.76 - 7 \cdot 4.3 = 6.7$ eV to be compared with an actual value close to 6.3 eV.

C. Comparison with Iodide, Xenon and Caesium

The electron affinity of the gaseous iodine atom is the ionization energy I_0 of the gaseous I^- determined (from the first ionization limit in the near ultra-violet (44) in the absorption spectrum of a vapour mixture containing appreciable amounts of this species) to be 3.06 eV. The first excited level $^2P_{\frac{1}{2}}$ of the iodine atom also belongs to the lowest configuration $[Kr]\ 4d^{10}\ 5s^2\ 5p^5$ and corresponds to a second ionization limit at 4.01 eV. One does not observe discrete energy levels of I^- though it would not be excluded that configurations terminating $5p^5\ 4f$ or $5p^5\ 6h$ would be just below the ionization limit. It is striking (45) that I^- in aqueous solution has two absorption bands at 5.5 and 6.4 eV which are only moving some 0.5 eV toward lower energy in acetonitrile, whereas cooled crystals of NaI, KI and RbI (46) show sharp absorption bands at 5.8 and 6.7 eV. The two transitions ostensibly separated by the same distance as the two J-levels of the iodine atom were ascribed by *Franck* and *Scheibe* to electron transfer to the solvent, a hypothesis which has persisted for a long time in literature. However, a more attractive description is Laporte-allowed inter-shell excitation to the configuration (neglecting closed shells) $5p^5\ 6s$, because at about 1 eV higher energy, other sharp absorption bands (46) of crystalline iodides correspond to $5p^5\ 5d$ in close analogy to the xenon atom having the four J-levels of $5p^5\ 6s$ 8.31, 8.43, 9.44 and 9.57 eV and the twelve J-levels of $5p^5\ 5d$ distributed between 9.89 and 11.61 eV to be compared with the two ionization limits $(5p^5)$ 12.13 and 13.44 eV of Xe.

When thin films of solidified xenon are studied, or a low concentration of xenon atoms are trapped in a cooled matrix (47,48) such as solid krypton or argon, the transitions to the excited configurations terminating $5p^5\ 6s$ and $5p^5\ 5d$ are seen as relatively narrow absorption bands, usually shifted some 0.5 to 1 eV toward higher wave-numbers, but 0,08 eV toward lower energy in crystalline xenon.

It is interesting to compare this evidence for localized inter-shell transitions of iodide and xenon in condensed matter with the behaviour of caesium. At sufficiently low pressure (49) the series $6s \rightarrow np$ can be observed in absorption with all the n values up to 73, converging to the ionization limit at 3.894 eV. However, this is not the last absorption lines of the caesium atom. *Beutler* and *Guggenheimer* (50) observed the two J-levels belonging to [Kr] $4d^{10} 5s^2 5p^5 6s^2$ at 12.31 and 13.52 eV and several J-levels of [Kr] $4d^{10} 5s^2 5p^5 5d 6s$ between 14.07 and 16.65 eV. It is worthwhile to compare these energies with Cs$^+$ (isoelectronic with Xe) having the four J-levels of [Kr] $4d^{10} 5s^2 5p^5 6s$ at 13,31, 13.37, 15.16 and 15.22 eV and the twelve J-levels of [Kr] $4d^{10} 5s^2 5p^5 5d$ between 13.17 and 15.33 eV. The shift about 1 eV of the first two auto-ionizing levels of Cs below the corresponding discrete levels of Cs$^+$ represent the binding of the external $6s$ electron by the increased attraction of the atomic core lacking a $5p$ electron. It is conceivable that the higher auto-ionizing levels of the caesium atom in the 15 to 17 eV region involve unusually strong configuration mixing.

Fig. 2 gives a semi-quantitative description of the optical excitations and ionization processes in iodides and xenon. The Madelung potential (calculated for spherically symmetric, non-overlapping ions) is calculated to be 7.8 eV in NaI and 6.4 eV in CsI with the result that the lowest ionization energy of iodide is expected to be 10.8 and 9.5 eV, respectively, to be compared with the I' values (corrected for the charging effects of the quasi-stationary positive potential acquired

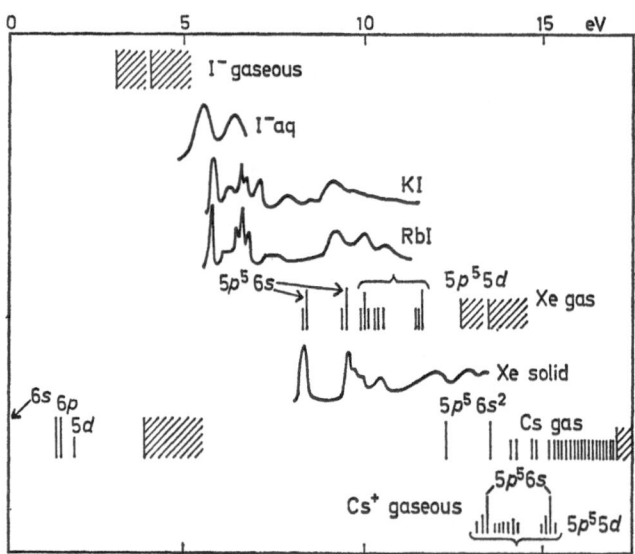

Fig. 2. Excited levels of iodide in gaseous state, aqueous solution and in cooled crystals of KI and RbI, of xenon in the gaseous and solid state, of caesium and of gaseous Cs$^+$. The excited levels accessible by symmetry-allowed transitions are given as higher lines, the continua are dashed

by bombardment of non-conductors with soft X-rays) 8.4 and 8.1 eV obtained from photo-electron spectra (51, 52). Said in other words, the 6s-like orbital known from the first absorption band (46) of these crystals is bound to the extent 2.4 eV to be compared with 3.82 eV in the first excited level of the xenon atom. The influence of the Madelung potential is the other way round on cationic sites, and the I' (Cs 5p) values observed (51, 52) are 15.0 eV in CsCl and CsBr and 15.1 eV in CsI. The absorption spectra (46) of crystalline caesium halides in the far ultra-violet show inter-shell transitions (found at much higher energy in rubidium and potassium halides) starting at 11 to 12 eV which possibly can be be identified with $5p \rightarrow 6s$ or $5p \rightarrow 5d$ localized on caesium(I).

Fig. 3 gives a comparable representation of internal $4f^q$ transitions and intershell $4f \rightarrow 5d$ and $4f \rightarrow 6s$ excitations of M^{+2}, M(II) in CaF_2, M^{+3}, MF_3 and MI_3 including electron transfer bands in the latter case. The distances between the

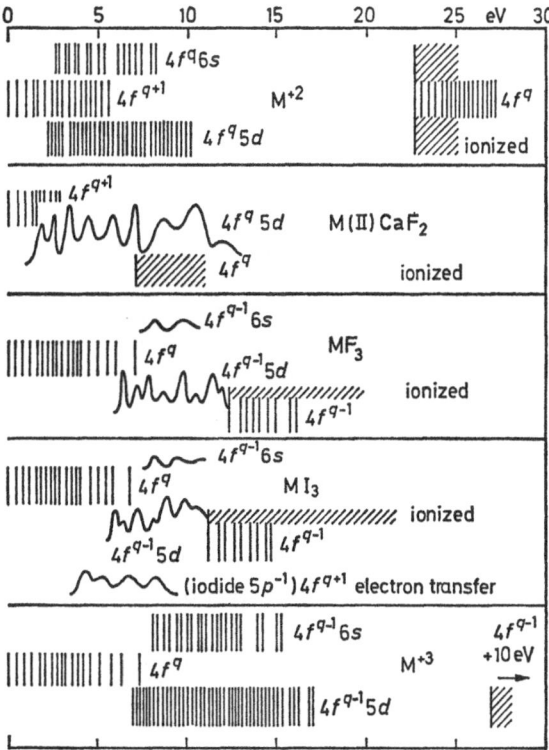

Fig. 3. Qualitative representation of excited levels of gaseous $4f$ group M^{+2} and M^{+3}, of M(II) in CaF_2 and of M(III) fluoride and iodide

thirteen J-levels of [Xe] $4f^2$ are known to decrease slightly by the *nephelauxetic* (cloud-expanding) *effect* (3, 40, 53) to the extent of some 4 percent from gaseous Pr^{+3} (54) to PrF_3 and some 2 percent going from PrF_3 to PrI_3. A comparison of $4f^3$ Nd(III) compounds shows smaller effects than in Pr(III), and the nephel-auxetic variation is less than half as pronounced in compounds of the heavier lanthanides from $4f^5$ Sm(III) to $4f^{12}$ Tm(III). Unfortunately, the published information about gaseous M^{+3} is very sporadic (2, 28). A very moderate nephel-auxetic effect is also perceived in Sm(II) in various crystals, but on the whole, broad absorption bands due to the low-lying $4f^{q-1} 5d$ cut off the $4f^q$ spectra in most M(II) with the result that only ζ_{4f} can be determined of the lowest term.

II. The Copenhagen Principle of Final States

A. Photo-electron Spectra of Metallic Lanthanides

It is possible to express various worries whether the text-book version of quantum mechanics is fully satisfactory, and *Ballentine* (55, 56) and the writer (57) have discussed whether it is only applicable to systems which are so small that they can be repeated in identical copies. However, this is the case for all monatomic entities and for all molecules and polyatomic ions of interest to the inorganic chemist. In this case, the sudden interaction of photons (one usually bombard gaseous samples (58) with 21.2 or 40.8 eV photons and solid samples (51, 52) with 1253.6 or 1486.6 eV photons) with a system has a variety of possible results, mainly consisting of an electron being ejected with the kinetic energy E_{kin} which is then interpreted as ionization needing the energy I defined as the difference between the photon energy and E_{kin}. Seen from this point of view, the system performs as a honest gambling-machine. When the slug (photon) is accepted, each outcome has a definite probability, and the choice between the alternative results has not at all a sufficient cause in the Medieval sense, any more than the actual time of decay of a radioactive nucleus. An analogy to radioactive isotopes (or strictly isomers) possessing competing modes of decay is that the half-life $t_{1/2}$ of the ionized system is the reciprocal value of the sum of the reciprocal values of effective half-lifes for various processes (such as ejection of an Auger electron leaving the system in another of various states, or differing photon energies of X-ray emission). According to the Heisenberg uncertainty principle, the one-sided half-width δ of the photo-electron signal is 2.3 eV divided by $t_{1/2}$ of the ionized system in the unit 10^{-16} sec. In actual practice, δ observed on the photo-electron spectra recorded (57) is the square-root of the sum of various squared contributions to the width, such as instrumental dispersion of E_{kin} and lack of monochromacy of the photons, and also vibrational broadening due to the Franck-Condon principle not allowing the internuclear distances to vary. In many instances, the Heisenberg uncertainty principle provides an insignificant contribution to δ if $t_{1/2}$ is well above 10^{-15} sec.

We are not going here to discuss the photo-electron signal satellites due to shake-up and shake-off processes, which are weak and most readily detected in gaseous atoms (59). They have, however, the connection with the principle of final states that the barycentre of all the photo-electron intensity including the main signal and all the satellites are situated at the *Koopmans* value of ionization energy, assuming "frozen orbitals" of the *Hartree-Fock* wave-function of the groundstate, whereas the main peak at lowest I corresponds to "relaxed" electronic density of the other orbitals (in monatomic entities, the radial functions contract). This *Manne-Åberg principle* (60) is an analogy to the Franck-Condon principle of invariant internuclear distances that the other orbitals, in a certain sense, do not have the time to adapt. but that the system lacking an electron collapses to a variety of Ψ being eigen-functions of the *new* Hamiltonian in agreement with the principle of final states. Normally, the largest probability occurs for the same electron configuration (say $1s\,2s^2\,2p^6$ of Ne^+) of the adapted electronic density, and the shake-up situation $1s\,2s^2\,2p^5\,3p$ has a much smaller probability, though it is important to note that the final states are eigen-functions of Ne^+. The *relaxation energy* (16) corresponding to the difference between the *Koopmans* value (890 eV in the case of neon $1s$) and the observed I (870.3 eV in the example) can be divided in an *intra-atomic* and an *inter-atomic* part. Empirically, the intra-atomic relaxation energy for inner shells is close to 0.8 eV times the square-root of I (in eV). Experimental evidence for inter-atomic relaxation effects can be obtained (61, 62) for neon atoms incorporated in metallic copper, silver and gold showing decreases of $I(Ne\,1s)$ between 3.5 and 4 eV. In molecules containing carbon, nitrogen and oxygen, the relaxation energy (63) is between 14 and 23 eV for the ionization of a $1s$ electron. The variation of the inter-atomic relaxation energy from one compound to another of the same element contributes a significant amount (52, 64) of the chemical shift dI of inner-shell ionization energies.

However, systems with partly filled $4f$ shells show a specific consequence of the principle of final states. *Wertheim, Rosencwaig, Cohen* and *Guggenheim* (65) and *Jørgensen* and *Berthou* (51) found independently that the ionization $4f^q \rightarrow 4f^{q-1}$ produces two signals in MF_3 and other non-conducting compounds in the elements from terbium ($q=8$) to ytterbium ($q=13$) whereas the elements from praseodymium ($q=2$) to gadolinium ($q=7$) show essentially one signal. This can be understood qualitatively as the possibility of changing the groundstate S to $(S+\frac{1}{2})$ *or* $(S-\frac{1}{2})$ in the second half of the $4f$ shell, whereas S can only decrease half a unit by ionization in the first half. The difference between the two barycenters in Eq. (5) is $(2S+1)\,D$, where the numerical value of $7D\sim5.6$ eV is in qualitative agreement with the observed separation 6.7 eV in terbium(III) compounds, whereas $2D$ is distinctly smaller than the distance 3.8 eV between the two signals of ytterbium(III) compounds.

This discrepancy is expected, because each S-barycentre represents a large number of J-levels. Actually, the configuration f^q consists of:

$q =$	0,14	1,13	2,12	3,11	4,10	5,9	6,8	7	
States	1	14	91	364	1001	2002	3003	3432	(8)
J-levels	1	2	13	41	107	198	295	369	
S,L-terms	1	1	7	17	47	73	119	119	

distributed over an interval to which $6D$ is a lower limit for $q=4$ and 10; $8D$ for $q=5$ and 9; $12D$ for $q=6$ and 8, and $15D$ a lower limit for $q=7$ because of Eq. (5). However, one does not observe photo-electron signals of $4f^{q-1}$ in lanthanides dispersed over such a wide interval, in particular only one signal of Gd(III). It would be rather hopeless to distinguish the individual photo-electron signals if their intensity was proportional to the number of states $(2J+1)$ in each level or $(2S+1)(2L+1)$ in each term. *Cox (66, 67)* proposed that the intensity of each signal is proportional to the *squared coefficient of fractional parentage* of the $4f^{q-1}$ level or term in the S, L, J ground level of $4f^q$ assuming spherical symmetry to remain a good approximation in the lanthanides. In the special case of $4f^{13}$ known from ytterbium(III) compounds, but not from metallic ytterbium being closed-shell $4f^{14}$ with the conditional oxidation state Yb[II] though it becomes Yb[III] at high pressure *(68)* or in globules with a diameter below 30 Å, as can be seen from X-ray emission and electron diffraction rings *(69)*, each term (though not each level) of the ionized $4f^{12}$ Yb[IV] has a probability of formation $(2S+1)$ $(2L+1)/7$ proportional to the number of states in each term. *Cox (66)* normalizes this probability to q when ionizing $4f^q$.

Table 1 contains these probabilities in all cases where a term is formed with higher normalized probability than 0.5. For a given value of S, there is a tendency toward proportionality to $(2L+1)$. For q at most 7, the only accessible terms of $4f^{q-1}$ have S decreased half unit. Since L can at most change 3 units by removal of a single f electron, f^3 cannot form ^3P, f^4 cannot form ^4S and ^4D, and f^5 cannot form ^5S (though ^5D has the probability 0.476 included in Table 1). When f^8 is ionized, ^8S is formed with $(8/7)$ and each of the six sextet terms of f^7 with the probability $(2L+1)/7$. For q above 7, the probability of ionizing to $(S+\frac{1}{2})$ is:

$$q = 8 \qquad 9 \qquad 10 \qquad 11 \qquad 12 \qquad 13 \qquad\qquad (9)$$
$$\quad \ ^{8}/_{7} \quad\ \ ^{14}/_{6} \quad\ \ ^{18}/_{5} \quad\ \ ^{20}/_{4} \quad\ \ ^{20}/_{3} \quad\ \ ^{18}/_{2}$$

with the result that the probability of ionizing to all the levels $(S-\frac{1}{2})$ is $8(14-q)/(15-q)$. Hence, when applicable, the relative probability $(S+\frac{1}{2})/(S-\frac{1}{2})$ is

$$(q-7)(16-q)/(112-8q) \qquad\qquad (10)$$

The squared coefficients given in Table 1 for ionization of f^9 to ^5L, ^5K and ^5H and of f^{12} to ^2L, ^2K and ^2H are all $(2L+1)/11$, and for ionization of f^{10} to ^4M and ^4L and of f^{11} to ^3M and ^3L are all $(2L+1)/13$. It is possible *(15)* to calculate diagonal and non-diagonal elements of interelectronic repulsion in the numerous cases where the configuration f^{q-1} presents two or more terms with the same combination of S and L. *Cox (66)* selected a definite choice of eigen-vectors reasonable close to the actual conditions. Certain sum rules are valid in these cases. Thus, the two ^4K formed when ionizing f^{10}, and the two ^3K formed from f^{11} have the sum $(15/13)$ as a single term with $L=7$ was expected to have. However, in many other cases, the sums are smaller than $(2L+1)/[(q-7)(14-q)+1]$.

Usually, the resolution obtained in photo-electron spectra is not sufficient to show individual J-levels, and the results given in Table 1 are indeed calculated for *Russell-Saunders* coupling. However, *Cox* has also calculated probabilities of formation of J-levels. With exception of the rather trivial case of $4f^{14}$ having the probability of 8 of forming $^2F_{7/2}$ (at lowest I) and 6 of $^2F_{5/2}$, the energetically

Table 1. Ionization energies in eV (I^* relative to the *Fermi* level) of metallic lanthanides and their antimonides, forming various final states (I_0^* the lowest) with probabilities calculated by *Cox* and assignments obtained by comparison with the levels of the M(III) aqua ion of the preceeding element

Groundstate	I^*(metal)	$I^*-I_0^*$	I^*(MSb)	$I^*-I_0^*$	Assignment	M(III)level		Probability
$4f^3$ $^4I_{9/2}$	Nd 4.8	0	5.8	0	$4f^2$ $^3H_4$	Pr 0	^3H	2.333
	(5.4)	0.6	(6.4)	0.6	$^3F_2$	0.6	^3F	0.667
$4f^5$ $^6H_{5/2}$	Sm 5.3	0	6.1	0	$4f^4$ $^5I_4$	Pm 0	^5I	2.758
	(6.8)	1.5	−	−	$^5F_1$	1.5	^5F	0.500
	7.6	2.3	8.5	2.4	$^5G_2$	2.2	^5G	1.266
	9.4	4.1	−	−	$^5D_0$	3.7	^5D	0.476
$4f^7$ $^8S_{7/2}$	Gd 8.0	−	9.1	−	$4f^6$ ^7F	Eu 0—0.6	^7F	7.000
$4f^8$ $^7F_6$	Tb 2.3	0	3.2	0	$4f^7$ ^8S	Gd 0	^8S	1.143
	7.4	5.1	8.2	5.0	^6I	4.5	^6I	1.857
	−	−	−	−	^6D	5.0	^6D	0.714
	9.3	7.0	10.1	6.9	^6G	6.2	^6G	1.286
	−	−	−	−	^6F	6.8	^6F	1.000
	10.3	8.0	11.1	7.9	^6H	7.3	^6H	1.571
$4f^9$ $^6H_{15/2}$	Dy 3.9	0	5.0	0	$4f^8$ $^7F_6$	Tb 0	^7F	2.333
	6.7	2.8	(8.3)	3.3	$^5D_4$	2.5	^5D	0.159
	7.7	3.8	8.7	3.7	$^5L_{10}$	3.4	^5L	1.545
	8.8	4.9	(9.7)	4.7	$^5H_7$	3.9	^5H	1.000
	9.3	5.4	10.3	5.3	$^5I_8$	4.4	^5I	0.919
	10.5	6.6	(11.5)	6.5	$^5K_9$	4.9	^5K	1.364
	12.5	8.6	13.5	8.5	?	−	−	
$4f^{10}$ $^5I_8$	Ho 5.3	0	6.0	0	$^6H_{15/2}$	Dy 0	^6H	2.800
	(6)	0.7	7.0	1.0	$^6F_{11/2}$	1.0	^6F	0.800
	−	−	−	−	$^4I_{15/2}$	2.7	^4I	0.636
	8.8	3.5	9.4	3.4	$^4M_{21/2}$	3.2	^4M	1.462
	(9.7)	4.5	(10)	4	$^4L_{19/2}$	4.0	^4L	1.308
	−	−	−	−	$^4K_{17/2}$	3.3	^4K	0.594
	−	−	−	−	−	−	^4K	0.559
$4f^{11}$ $^4I_{15/2}$	Er 4.8	0	5.6	0	$^5I_8$	Ho 0	^5I	3.182
	5.4	0.6	(6.3)	0.7	$^5I_7$	0.6	−	−
	6.8	2.0	(8)	2.4	$^5F_5$	1.9	^5F	0.625
	−	−	−	−	$^3K_8$	2.6	^3K	0.594
	7.7	2.9	8.7	3.1	$^5G_6$	2.7	^5G	1.193
	8.7	3.9	9.6	4.0	$^3L_9$	3.6	^3L	1.308
	9.4	4.6	10.2	4.6	$^3M_{10}$	4.3	^3M	1.462
$4f^{12}$ $^3H_6$	Tm 4.6	0	5.5	0	$^4I_{15/2}$	Er 0	^4I	3.677
	5.8	1.2	6.5	1.0	$^4I_{13/2}$	0.8	−	−
	−	−	−	−	$^4F_{9/2}$	1.9	^4F	0.677
	(7)	2.4	8.2	2.7	$^2H_{11/2}$	2.3	^2H	1.000
	8.3	3.7	9.3	3.8	$^4G_{11/2}$	3.3	^4G	1.688
	−	−	−	−	$^2K_{15/2}$	3.5	^2K	1.364
	10.1	5.5	11.2	5.7	$^2L_{17/2}$	5.2	^2L	1.545

lowest J-level of a given term (having $J=|L-S|$ for q below 7 and $J=L+S$ for q above 7) takes over nearly all the intensity belonging to the term. Thus, $^3H_4$ of f^2 has the probability (12/7) of forming $^2F_{5/2}$ and only (2/7) of $^2F_{7/2}$ whereas $^2F_{7/2}$ of f^{13} distributes the probability $^3H_6$ (3.018), $^3H_5$ (1.375) and $^3H_4$ (0,321). This situation can be even more extreme in the previous heavy lanthanides. Thus, $^3H_6$ of f^{12} has the probability 1.500 of forming $^2L_{17/2}$ but only 0.045 of forming $^2L_{15/2}$.

Hagström and collaborators (*70, 71*) started studies of photo-electron spectra of metallic lanthanides. This is a considerable experimental problem, needing pressures below 10^{-10} torr and freshly evaporated samples. These investigations have been continued by *Baer* (*66, 72*) recently using monochromatized 1486.6 eV photons, achieving the unusually good resolution 0.3 eV. In Table 1 are given the I^* values relative to the Fermi level. The identification is performed by comparison with the absorption spectra of $4f^{q-1}$ of the M(III) aqua ions of the previous element (*73*) and of Gd(III) in CaF_2 (*74*). This assignment of J-levels was previously discussed (*6*) but the more complete list given by *Carnall, Fields* and *Rajnak* (*73*) contains the high L-values combined with the next-highest S which are particularly important for photo-electron spectra because so high ionization probabilities are concentrated in a term (and, specifically, in its lowest J-level).

The agreement has fully confirmed the original idea of *Cox, Evans* and *Orchard* (*75*) that the photo-electron signals have intensities proportional to the squares of the coefficients of fractional parentage. However, the detailed distribution, and in particular the distance $(I^* - I_0^*)$ above the lowest $4f^{q-1}$ signal, contains rather interesting information. It is not surprising that the levels of M[IV] obtained by photo-ionization have a slightly wider distribution than the isoelectronic M(III) of the preceeding element. Actually, gaseous M^{+4} is expected (*3, 6, 53*) to have term distances 20 percent higher than the isoelectronic M^{+3} [for comparison, it may be noted that the parameters of interelectronic repulsion increase linearly some 3 percent per unit of Z from $4f^2$ Pr(III) to $4f^{12}$ Tm(III)]. Though the quantity 20 percent has not been verified by atomic spectroscopy, a comparison with the other transition groups would make it improbable that it is outside the interval 17 to 22 percent. Hence, the increase of term distances 12 percent found (*66*) going from Gd(III) $4f^7$ to metallic Tb, and from $4f^8$ Tb(III) to Dy, or the more moderate increase 6 percent from $4f^{11}$ Er(III) to Tm indicate a *nephelauxetic ratio* $\beta = 0.92$ for Tb[IV] and Dy[IV] and 0.87 for Tm[IV] taking into account the weak nephelauxetic effect ($\beta = 0.98$ or 0.99) already occurrring in the M(III) aqua ion compared with M^{+3}. It is not unexpected that terbium(IV) forming many mixed oxides, and dysprosium(IV) known from the fluoride $Cs_3 Dy F_7$ show a less pronounced nephelauxetic effect than thulium(IV) too oxidizing to form any compounds. However, a value as high as $\beta = 0.87$ (or perhaps 0.84 in ytterbium(III) photo-electron spectra) is higher than $\beta = 0.79$ for $Cr(H_2O)_6^{+3}$ and 0.76 for $Fe(H_2O)_6^{+3}$ though chromium(III) and iron(III) are far less chemically oxidizing. We return to this dilemma below.

The confirmation of the principle of final states (though combined with the *Franck-Condon* principle of invariant internuclear distances) by the photo-electron spectra of metallic lanthanides invites the question where the satellite signals needed to obey the *Manne-Åberg* principle (*60*) occur. There is a mildly undulating background at higher I^* before the $5p$ region, but a specific case may be the unidentified signal of metallic dysprosium (72) at $I^* = 12.5$ eV. Whereas it is not probable that shake-up to $4f^{q-2} 5d$ is important, because the expected I^* are in the upper end of the regions studied (in particular in metallic ytterbium with $I^* = 1.1$ and 2.4 eV and in lutetium with $I^* = 7.1$ and 8.6 eV due to $^2F_{7/2}$ and $^2F_{5/2}$ of $4f^{13}$), it is conceivable that dysprosium shows shake-off to $4f^7$ loosing two electrons simultaneously, and the sharp signal being favoured by the isolated ^8S state. An alternative, and more pedestrian, explanation is that a triplet level of $4f^8$ Dy[IV] has acquired sufficient quintet character by effects of intermediate coupling, though this seems less likely when comparing the distance 2.0 eV from the neighbour signal with $\zeta_{4f} = 0.2$ eV.

The highest $I^* = 8.0$ eV for the first signal (found in gadolinium) indicates a lower limit to the electron affinity in view of the obvious fact that the conduction electrons do not invade the $4f$ shell. *Herbst, Lowy* and *Watson (76)* and *Hüfner* and *Wertheim (77)* estimate the difference between the ionization energy and the electron affinity in metals to be 7 eV. In condensed matter, it is expected that this difference is much smaller than the difference (35) between $I_3 = 20.4$ and $I_4 = 44.0$ eV of Gd^{+2} and Gd^{+3} but a comparison (51) of the photo-electron spectra of terbium(III) and terbium(IV) in a mixed oxide suggest a difference 16 eV of I, of which 7 eV is due to the differing influence of spin-pairing energy from Eq. (5) and 9 eV represents the one-electron contribution. Tb^{+3} has $I_4 = 39.8$ eV (35). Metallic Gd and Tb have I^* 36.0 and 37.5 eV lower than I_4, respectively. It may be noted that the linear increase of $I(4f_{7/2})$ between hafnium ($Z = 72$) and gold ($Z = 79$) 9 eV per unit of $Z(20,51)$ still is only 5 times the hydrogenic value 27.2 eV/$n^2 = 1.7$ eV.

B. Photo-electron Spectra of Lanthanide Compounds

We already mentioned the studies (20, 51, 65) of $4f$ group fluorides and oxides originally found to be in agreement with the refined spin-pairing energy theory. In Table 2 are summarized the I' values for such compounds, corrected (51, 52) for the influence of the quasi-stationary positive potential maintained by non-conductors in the photo-electron spectrometer.

Unfortunately, the resolution is not as good as when a monochromatized photon beam, and ultra-high vacuo are applied. The first class of $4f$ group compounds which have been measured under such conditions are the antimonides MSb crystallizing in the NaCl type, of which *Campagna, Bucher, Wertheim, Buchanan* and *Longinotti (78)* produced freshly cleaved surfaces inside the instrument. These authors give the I^* values relative to the Fermi level. All the samples, including LaSb, show a signal close to $I^* = 2$ eV essentially due to Sb $5p$ electrons. The resolution of the $4f$ signals is nearly as good as in the metallic elements. As seen in Table 1, the same structure is repeated at about 0.9 eV higher I^*.

Using the same technique, *Campagna, Bucher, Wertheim, Buchanan* and *Longinotti (79)* provided rather amazing evidence for the Copenhagen principle of final states in two highly unusual compounds Tm Te and Tm Se. The crystals MS, MSe and MTe all crystallize in NaCl type, and are semi-conducting M(II) compounds for M = Eu and Yb but metallic and showing the magnetic properties

Table 2. The lowest ionization energy $4f^q \rightarrow 4f^{q-1}$ compared with calculated values (*33*) of I_{chem} for M(III) aqua ions and I_4 of gaseous M^{+3} (*35*) all in eV

$q =$		M $I^* + 3$ eV	MSb $I^* + 3$ eV	M_2O_3 I'	MF_3 I'	I_{chem} M(III)	I_4 M^{+3}
2	Pr	6.3	7.6	9	10.7	7.9	38.98
3	Nd	7.8	8.8	10.4	12.2	9.1	40.41
5	Sm	8.3	9.1	11.1	12.8	9.7	41.37
6	Eu	—	—	12.0	13.7	10.9	42.65
7	Gd	11.0	12.1	14.1	15.5	12.4	44.01
8	Tb	5.3	6.2	8.5	9.0	7.8	39.79
9	Dy	6.9	8.0	9	10	9.5	41.47
10	Ho	8.3	9.0	10	12	10.7	42.48
11	Er	7.8	8.6	9	11.9	10.6	42.65
12	Tm	7.6	8.5	10	—	10.6	42.69
13	Yb	—	—	12.5	14.0	11.6	43.74
14	Lu	10.1	—	13.9	15.3	13.0	45.19

of M[III] for M = Ce, Pr, Nd, Sm, Gd, Tb, Dy, Ho and Er. The condition for metallicity (*80, 81*) seems to be that the groundstate would be $4f^{q-1}$ 5d rather than $4f^q$. Seen from this point of view, M = Tm represents an interesting border-line case. The magnetic susceptibility of TmTe is intermediate between the values expected for Tm(II) and Tm(III). As seen on Fig. 4, the photo-electron spectrum of TmTe shows a superposition of the $4f^{12} \rightarrow 4f^{11}$ signals at higher I^* (also known from metallic thulium and from TmSb) and a set of $4f^{13} \rightarrow 4f^{12}$ signals of comparable intensity at lower I^*. Hence, on an instantaneous picture (*19*) this cubic crystal

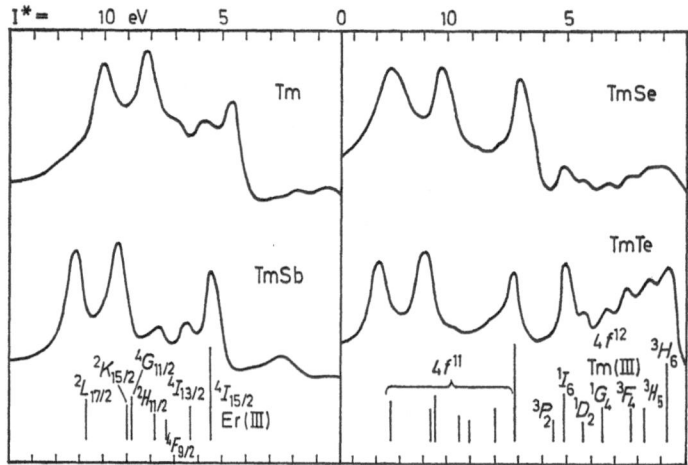

Fig. 4. Photo-electron spectra (I^* relative to Fermi level) of metallic thulium, and thulium antimonide, selenide and telluride. The $4f^{11}$ energy levels of Er(III) and $4f^{12}$ energy levels of Tm(III) aqua ions are given for comparison, with the height of the lines determined by the probability calculated by Cox

contains about equal quantities of Tm[III] and Tm[II] though a "hopping" of electrons from the $4f^{13}$ sites to $4f^{12}$ sites probably takes very short time. The isotypic selenide TmSe contains about 80 percent Tm[III] and 20 percent Tm[II] on an instantaneous picture.

Perhaps the most interesting aspect of this surprising result is that the final states $4f^{11}$ Tm[IV] formed by ionization of the $4f^{12}$ systems have a nephelauxetic effect comparable to TmSb in Table 1 but that the $4f^{12}$ Tm[III] formed from $4f^{13}$ (on crystallographically equivalent sites) hardly shows any nephelauxetic effect compared with the energy levels of thulium(III) aqua ions (73, 82):

	$^3H_6$	$^3F_4$	$^3H_4, ^3F_3$	$^1D_2$	$^1I_6$	
TmTe, $(I^* - I_0^*)$:	0	0.7	1.6	3.4	4.3	(11)
Tm(III) aqua ion:	0	0.7	1.5, 1.8	3.5	4.35	

By the way, this is the first instance of well-resolved $4f^{13} \rightarrow 4f^{12}$ photo-electron signals; the ten non-conducting ytterbium(III) compounds (16, 51) studied show less resolution. One might hope to study an ytterbium alloy containing Yb[III].

As seen in Table 1, there is marginal evidence for stronger nephelauxetic effect in Tb[IV] and Dy[IV] formed by ionizing TbSb and DySb than the elements, whereas Tm[IV] formed in TmSb has a *smaller* nephelauxetic effect than in metallic thulium. The energy difference 5.2 eV between $^2L_{17/2}$ and the groundstate $^4I_{15/2}$ of Er(III) has increased to 5.5 eV in Tm, 5.7 eV in TmTe and TmSb, and is expected to be 6.3 eV in gaseous Tm^{+4}.

Table 2 also gives $I_{chem} = E^0 + 4.5$ eV according to Eq. (7) derived from the standard oxidation potentials E^0 of $4f$ group M(III) aqua ions calculated by *Nugent, Baybarz, Burnett* and *Ryan* (33) partly by extrapolation of the electron transfer spectra of various types of M(IV) halide complexes. It is striking that these I_{chem} values (not restricted by the *Franck-Condon* principle) agree with the refined spin-pairing energy theory by all being 1.3 to 3 eV higher than $(I_0^* + 3$ eV$)$ for the metallic elements, 0,3 to 2 eV higher than $(I_0^* + 3$ eV$)$ for MSb, within 1 eV from the I' values of $M_2 O_3$ and some 2 to 3 eV *lower* than I' of MF_3. For comparison, Table 2 also gives I_4 of gaseous M^{+3} (35) being 31.1 to 32.2 eV higher than I_{chem}. This difference is 7 times the hydration energy parameter $\varkappa$ (3, 43) seemingly varying between 4.44 and 4.6 eV. Though higher than the accepted value 4.3 eV, this is not unreasonable in view of the smaller ionic radii of the chemically unknown M(IV) species.

The photo-electron spectra suggest the lowest ionization energy $4f^q \rightarrow 4f^{q-1}$ to vary along the double zigzag curve known (Fig. 1) from the other system differences, with a *chemical shift* some 4 to 5 eV from the metallic element to the fluoride and with the antimonides, oxides and iodates (51) at intermediate positions.

C. Why is the Number of $4f$ Electrons an Integer?

The answer to this question has several parts, among which one aspect is already present in monatomic entities. Suppose two levels with the same combination of J, S and L of Pr^+ belonging to the two configurations $[Xe] 4f^2 6s^2$ and

67

[Xe] $4f^4$ have a distance comparable to twice their non-diagonal element of interelectronic repulsion (perhaps 0.2 eV) their eigen-vectors contain about equal amounts of squared amplitude of the two configurations. However, one would not normally say that three $4f$ electrons occur in such a situation. It is true that one can make a kind of *Mulliken* population analysis of such a monatomic entity, and in the case of two electrons outside closed shells, the higher L values have the higher population numbers of the f^2 shell for the simple reason that s^2 can mix only with ^1S, p^2 can mix with ^1S, ^3P and ^1D, d^2 with these three terms and ^3F and ^1G, whereas ^3H and ^1I still can be influenced by g^2, h^2, ... When a chemist asks the question of the Chapter heading, he is not so much interested in this problem connected with the behaviour of natural spin orbitals (6, 83) but rather in a much more familiar context which can be described with the LCAO. approximation.

As seen on Fig. 3, there is no doubt from a classificatory point of view that a certain number of levels of a given M(III) compound belong to [Xe] $4f^q$ followed at a certain distance by levels belonging either to [Xe] $4f^{q-1} 5d$ or to an electron transfer configuration (ligands)$^{-1}$ [Xe] $4f^{q+1}$. In the case of M(II) compounds, the first excited configuration is always [Xe] $4f^{q-1} 5d$, whereas it always is of the electron transfer type in M(IV). Regarding the manifold of adjacent levels, one cannot argue that any doubt adheres to the integer q. This statement would not be so clear-cut in the d groups (3, 18) where the five d-like orbitals can show rather different behaviour, for instance of delocalization on the ligating atoms. Another, related problem is catenation. It can be argued that the dimeric acetate $M_2(O_2CCH_3)_2$, $2 H_2O$ contain two $3d^3$ Cr[III] and $4d^6$ Rh[III] in the cases of M = Cr and Rh, where an additional electron pair (which may be d-like or not) binds the two M atoms, like $3d^6$ Mn[I] in $(OC)_5Mn$ $Mn(CO)_5$. Hence it is not certain whether these compounds have the oxidation states Cr(II), Rh(II) and Mn(0), or perhaps no oxidation state at all (3). Nevertheless, the unusual behaviour of thulium telluride (79) has opened up a new field of complications, though the choice between $4f^{12}$ Tm[III] and $4f^{13}$ Tm[II] is made by the atoms in a given instant.

After reviewing the evidence for q invariantly being an integer defining the conditional oxidation states M[II], M[III] and M[IV], we may discuss the physical mechanism. A major reason is the coefficient $q(q-1)/2$ to $A_*(4f, 4f)$ in Eq. (2) which would increase in direction of the value $q^2/2$ for a classical charge distribution if the $4f$ electrons formed an energy band in the sense of the conduction electrons of the alkaline and the coinage metals. As discussed in the next Chapters, the typical molecule has MO being in between these two extremes, loosing perhaps $-qA_*/4$ relative to the isolated atoms. The stabilization $-qA_*(4f, 4f)/2$ of monatomic $4f^q$ (due to the absence of self-repulsion in a given electron) is a considerable quantity. Since Dr. *Watson* kindly indicated $\langle r^{-1} \rangle = 1.61$ bohr^{-1} for his Hartree-Fock $4f$ orbitals of Gd^{+3}, $A_*(4f, 4f)$ considered as an integral of interelectronic repulsion close to $0.6 \langle r^{-1} \rangle$ atomic units is 27 eV. The observed (35) differences $(I_4 - I_3)$ corrected for spin-pairing energy give the phenomenological parameter 19 eV corresponding to the general Watson effect of dielectric diminution of the interelectronic repulsion (6, 84). It was discussed above how this difference is evaluated to 9 eV from the photo-electron spectra of Tb(IV) and

Hf(IV) compared with the isoelectronic Gd(III) and Lu(III) (20, 51) and estimated theoretically (76) to 7 eV for the metallic elements. *Campagna et al.* (79) determined the distance 6.4 eV between the lowest $I^*(4f)$ of the $4f^{12}$ and $4f^{13}$ sites in Tm Te and 5.6 eV in Tm Se giving $A_*(4f, 4f)$ about 0.4 eV larger values after correction for the quantities entering the refined spin-pairing energy theory.

Another correction favouring localized orbitals with high D is the spin-pairing energy Eq. (5). For Gd(III) the effective value of $-7 A_*(4f, 4f)/2$ in condensed matter is -28 eV (but -67 eV for gaseous Gd^{+3}) to be compared with $-168 D/13$ $= -11$ eV only expected to differ by one or two percent in the two cases. *Mott* (85) and *Hubbard* (86) call the activation energy for electron transfer from l^q to l^{q+1} the parameter U, which is essentially our A_* but corrected for the various other effects. It may be noted that the difference 12.85 eV between the ionization energy 13.60 eV and the electron affinity 0.75 eV of the hydrogen atom is smaller than $A_*(1s, 1s) = 1.25$ rydberg $= 17.0$ eV calculated for H, but larger than the Slater value $17.0(1 - 0.3125) = 11.7$ eV for H$^-$. Though of less chemical relevance, it is also interesting to compare $I_1 = 24.587$ and $I_2 = 54.416$ eV of helium with $17.0(2 - 0.3125) = 28.7$ eV, slightly smaller than $I_2 - I_1 = 29.83$ eV. This discrepancy is mainly due to correlation energy in the two-electron system In many-electron systems such as Gd^{+3}, and *a fortiori* in Gd(III) many additional effects (76) operate to decrease the phenomenological A_*.

III. Partly Covalent Bonding

A. The Variation Principle and the LCAO Model

The MO configuration $(\sigma_g)^2$ of the lowest state of the hydrogen molecule is approximately as well-defined as $1s^2$ of the helium atom, as far goes natural spin orbitals (87), and at the *equilibrium* internuclear distance, its correlation energy is about -1 eV (like He) though this is more than a-fifth of the dissociation energy of H$_2$. However, at larger R, the correlation energy decreases dramatically. The asymptotic behaviour for very large R corresponds to equal amplitude of exactly two MO configurations $(\sigma_u)^2$ and $(\sigma_g)^2$. The diagonal element of energy of both these configurations is the average of the combination H $+$ H (at which we put the zero-point of energy) and H$^+$ $+$ H$^-$, or 6.42 eV. The non-diagonal element consists (6) of the two-electron quantity $\frac{1}{2}A_*(1s, 1s) = K(\sigma_g, \sigma_u)$ for large R. If one neglects the overlap integral between the $1s$ orbitals of the two hydrogen atoms (this is always somewhat dangerous when discussing chemical bonding) the energy levels at large R can be written

$$(E_a + E_b) - (E_a - E_b)^2 / A_*(1s, 1s) \tag{12}$$

where E_a is the one-electron energy of the anti-bonding MO of symmetry type σ_u (having a node-plane between the two nuclei) and E_b of the bonding σ_g. The level given by Eq. (12) corresponds to a mixture of some 49 percent $(\sigma_u)^2$ and some 51 percent $(\sigma_g)^2$ and has $S = 0$. It is slightly below a level with $S = 1$ containing three states, and at large R having the exact MO configuration $(\sigma_g)^1$

$(\sigma_u)^1$ and the energy $(E_a + E_b)$. If R is decreased to such a value that $(E_a - E_b) = 0.5$ eV, Eq. (12) is still only -0.019 eV because the denominator is 13 eV. In addition to these four states, two other states formed by the distribution of two electrons on two orbitals $(4 \cdot 3/2 = 6)$ have much higher energy corresponding to linear combinations (having $S = 0$) of the situations $H^- H^+$ and $H^+ H^-$. The rapid separation of one of these six states as the groundstate of H_2 for smaller R corresponds to what *Moffitt* (88) calls the transition from Λ,S-coupling to valency-coupling.

It is important to note that the small stabilization of the singlet relative to the triplet given by Eq. (12) for large R as a second-order expression of the one-electron energy difference $(E_a - E_b)$ is the mechanism of *anti-ferromagnetic coupling*. *Heisenberg* introduced a parameter J producing energy levels of the form $+JS(S+1)$ where J is chosen positive for anti-ferromagnetic, and negative for ferromagnetic cases. With this definition, Eq. (12) indicates $2J$. The behaviour of a large number N of ions each having an individual value S_0 has been discussed (19) and in particular it has been shown (for vanishing J) that the average value of $S(S+1)$ for such a system is $NS_0(S_0+1)$. Since the quantity measured by magnetic susceptibility is proportional to $S(S+1)$ this explains why N ions are N times as paramagnetic as one ion. In the case of many ions, one should not forget the warning by *Griffith* (89) that the dispersion of energy levels having the same S may be rather large compared with J; the *Heisenberg* formula applies to the barycentre of all states having a given S. Though the interelectronic repulsion in Eq. (5) also depends on $S(S+1)$ (though in a monatomic system) the physical origin of anti-ferromagnetic coupling is entirely different. *Glerup* (90) discusses the dimeric blue $(NH_3)_5CrOCr(NH_3)_5^{+4}$ having the unusually large $J = 0.028$ eV finding the Heisenberg interval rule for the low-lying levels with $S = 0, 1, 2$ and 3 where the highest level with $S = 3$ is not stabilized, whereas the lower S-values are the lowest eigen-values of the type Eq. (12) of secular determinants having as diagonal elements states with a definite number of d-like electrons (2, 3, or 4) and correspondingly well-defined oxidation states, the lowest being Cr(III), Cr(III) and a higher set being Cr(II), Cr(IV) and Cr(IV), Cr(II). In the case of half-filled sub-shells and other *Kamimura*-stable systems (19, 91) one obtains the Heisenberg formula. The protonated pink $(NH_3)_5CrOHCr(NH_3)_5^{+5}$ is not linear $(< CrOCr = 165°.6)$ and has (92) a much smaller $J = 0.002$ eV. The more complicated ethylenediamine complex $Cr_4(OH)_6en_6^{+6}$ (93) has magnetic properties suggesting the four Cr(III) in a rhomb, confirming the independent crystal structure (94) containing two Cr(III)O_2N_4 and two Cr(III)O_4N_2 chromophores. The 256 low-lying states are calculated (93) to be distributed on 44 different levels having $S = 0, 1, 2, 3, 4, 5$ and 6 and energies being multiples of J from zero to $21J$. Solid-state physicists (95, 96) have elaborated models of anti-ferromagnetism rather foreign to chemical thinking. In the case of $4f$ groups compounds, one has to cool to very low temperatures before seeing deviations from Curie paramagnetism, and only in rather special mixed oxides (97) such as DyFeO$_3$ the *Heisenberg* parameter J achieves values above 10^{-4} eV.

Returning to molecules or polyatomic ions containing at most one partly filled shell, it is generally assumed that the MO (hopefully classifying the energy levels like the nl-shells of monatomic entities) have the character of *linear com-*

binations of atomic orbitals (LCAO). Other approximations have sometimes been used in crystals with repeating unit cells in translational symmetry, such as the "augmented plane-wave method" where the one-electron functions are supposed to be those of freely moving electrons in a constant potential, modified by orthogonalization on the closed shells of the relatively unvoluminous atomic cores. However, in recent years, the *Multiple-Scattering* $X\alpha$ *Method* proposed by *Johnson* and *Slater (98, 99)* represents an attempt of solving *Schrödinger's* equation directly without the LCAO assumption.

Nevertheless, the typical heteropolar chromophore MX_N is usually described by LCAO variants, among which the *Wolfsberg-Helmholz* model *(100, 101)* was very popular around 1960. A closer analysis *(6)* shows that the eigen-value corresponding to the anti-bonding orbitals of the partly filled shell is related to the diagonal elements of energy H_M and H_X (both negative) of the central atom and of the ligand orbitals, and to the overlap integral S_{MX} between the central atom orbital and the linear combination of ligand orbitals having the same symmetry (in the point-group of the chromophore) as the central atom orbital and the resulting MO eigen-vector, by the approximate expression

$$E_a = H_M + (H_X S_{MX})^2 / (H_M - H_X) \tag{13}$$

The eigen-value of the corresponding bonding orbital *(6)* is then

$$E_b = H_X - (H_M S_{MX})^2 / (H_M - H_X) \tag{14}$$

This is a reasonable description in heteronuclear molecules including the d-like electrons of VCl_4 *(102)* and various gaseous chromium(III) complexes *(103)* having lower ionization energies I than the filled MO which can be observed in photo-electron spectra of $TiCl_4$ and of the corresponding scandium(III) complexes. *Evans, Hamnett, Orchard* and *Lloyd (103)* pointed out that the iron(III) and cobalt(III) tris(hexafluoroacetylacetonates) have the rather alarming property of having higher I of the $3d$-like electrons than of the loosest bound MO mainly located on the ligands. This is less surprising in the case of the diamagnetic $(S = 0)$ Co(III) having all six d-like electrons in the lower, roughly non-bonding sub-shell (with angular functions proportional to xy, xz and yz avoiding the six oxygen atoms in the octahedral chromophore) than in the high-spin $(S = 5/2)$ iron(III) complex having one electron in each of the five d-like orbitals with the amazing corollary that the two anti-bonding electrons in the higher sub-shell has *higher* I than their bonding MO counterparts with the same symmetry type.

Since the present volume is about rare earths, there is no doubt that the same paradoxical situation occurs in nearly all solid $4f$ group compounds *(51, 52, 65)* with exception of Ce(III), Pr(III) and Tb(III). A unique, extreme situation occurs in $4f^7$ terbium(IV) *(52)* where $I'(4f) = 24.7$ eV in TbO_2 (kindly provided by Prof. *G. Brauer*) is somewhat higher than 22.4 eV in HfO_2. Before $5g$ group elements with Z above 121 are synthesized *(104)* terbium(IV) represents the highest I conceivable of a partly filled shell in chemical compounds.

It had been recognized for some time *(6)* that the electrostatic model of "ligand field" theory encounters almost as many difficulties in the $4f$ group as in the three

d groups, and the *angular overlap model* was originally introduced (*105*) with the purpose of explaining the spreading (usually 0.05 to 0.1 eV) of the one-electron energies of the seven $4f$ orbitals by anti-bonding effects proportional to the square of the overlap integral S_{MX} in expressions of the type Eq. (13). The angular overlap model can also be applied to partly filled d shells and was later shown (*19, 106–110*) to be equi-consequential with a contact potential acting at the positions of the ligand nuclei. Anyhow, in 1963, it was presumed (*105*) that the eigen-value E_a of Eq. (13) represents an ionization energy (we return to this rather intricate question in the next Chapter) and in the aqua ions $M(H_2O)_9^{+3}$ containing a chromophore $M(III)O_9$ of symmetry D_{3h} it was assumed (using evidence from electron transfer spectra) that H_M is about -6 eV, and H_X about -12 eV like in many oxygen-containing molecules. Whereas a few metallo-organic compounds (*52, 58*) of $3d^6$ chromium(0) have $I(3d)$ close to 6 eV, it cannot be seriously maintained that $E_a \sim -6$ eV should represent $I(4f)$ when Table 2 is known.

The realization that $I(3d)$ or $I(4f)$ can be higher than I of the loosest bound MO is colloquially called the *third revolution in "ligand field" theory* (*111, 112*) though it had previously been suspected (*6*) that such a situation might occur in cases such as $CuBr_4^{-2}$ or OsI_6^{-2} combining an oxidizing central atom with strongly reducing ligands. In many ways, this surprise was the least expected in the lanthanides, where the nephelauxetic ratio β is above 0.93 in $4f^2$ praseodymium(III) compounds (*2, 3, 19*) and probably above 0.95 in all other M(III) cases, and where the Landé parameter ζ_{4f} is much less sensitive to chemical bonding, in particular in $4f^{13}$ ytterbium(III) compounds. It cannot be argued that S_{MX} is below 0.03 explaining the weak covalent bonding, because one can always find a case where $I(M\,4f)$ coincides with $I(O2p)$ or I of other penultimate MO with the result that the ionized system has at least two degenerate eigen-values, of which it is known that any arbitrary linear combinations of these eigen-states is also a solution to the *Schrödinger* equation. It is not much better that all or most Eu(III), Gd(III) and Yb(III) compounds have the lowest I(4f) slightly above the lowest I of the ligands. Hence, it is not trivial that the parameters of the angular overlap model (*105, 113*) corresponds to anti-bonding effects, though the denominator in Eq. (13) seems to be negative (if represented by $I_M - I_X$) and that these parameters conceivably might have had the opposite sign of the d-group complexes. Further on, the anti-bonding effects decrease smoothly (*105*) from $4f^1$ Ce(III) to $4f^{13}$ Yb(III).

The way out of this labyrinth may be to consider (*112*) the total energy including the interelectronic repulsion. We construct two orbitals from ψ_M and ψ_X with the trigonometric parameter φ

$$\psi_a = (\cos\varphi)\,\psi_M - (\sin\varphi)\,\psi_X$$
$$\psi_b = (\sin\varphi)\,\psi_M + (\cos\varphi)\,\psi_X \tag{15}$$

These LCAO are written as if ψ_M and ψ_X have been orthogonalized (*19*). Let us assume that the system containing closed shells and one electron in the a orbital has its energy minimized for $\varphi = 20°$ (corresponding to significant covalent bonding) with $\sin\varphi = 0.34$ and $\cos\varphi = 0.94$. Then, the stabilization by covalency is tg $\varphi = 0.364$ times the non-diagonal element of the effective one-electron operator. In the configuration b^2, the set of coefficients to the parameters of interelectronic repulsion is

$$(\sin^4\varphi)\, A_*(M, M) + (2 \sin^2\varphi \cos^2\varphi)\, A_*(M, X) + (\cos^4\varphi)\, A_*(X, X) \qquad (16)$$

with the numerical values 0.01, 0.20 and 0.79 in the example. If $A_*(M, X) \simeq A_*(X, X)$ it is evident that the optimal value of φ only is modified to an insignificant extent. This is not at all true for the configuration $b^2 a^1$

$$[\sin^2\varphi(1 + \cos^2\varphi)]\, A_*(M, M) + [2 - 2 \sin^2\varphi \cos^2\varphi]\, A_*(M, X)$$
$$+ [\cos^2\varphi(1 + \sin^2\varphi)]\, A_*(X, X) \qquad (17)$$

with the numerical values 0.220, 1.794 and 0.986 of the three coefficients having the invariant sum 3. Here, the coefficient to $A_*(M, M)$ varies quite a lot and is:

$$\varphi = 15° \qquad 20° \qquad 25° \qquad 30° \qquad 45° \qquad 60° \qquad (18)$$
$$0.130 \qquad 0.220 \qquad 0.325 \qquad 0.4375 \qquad 0.75 \qquad 0.9375$$

When the total energy containing the contribution Eq. (17) is minimized, it is clear that a large value of $A_*(M, M)$ pulls the optimal φ down in direction of $15°$ or perhaps smaller values. In the $4f$ group, the non-diagonal element of the effective one-electron operator is perhaps only $0.04\, A_*(4f, 4f)$. In such a case, one may write the asymptotic expression for small φ, assuming $A_*(M, X) = A_*(X, X) = A$:

$$(2 \sin^2\varphi)\, A_*(M, M) + (3 - 2 \sin^2\varphi)\, A + (1 + \sin^2\varphi)\, H_M + (2 - \sin^2\varphi)\, H_X$$
$$+ (tg\varphi)\, H_{MX} \qquad (19)$$

When differentiating with respect to $\sin\varphi$, one finds to first order the minimum at

$$\sin\varphi = -\, (H_{MX})\, /\, [4\, A_*(M, M) - 4\, A + 2\, (H_M - H_X)] \qquad (20)$$

For the case $H_X = H_M$, this minimum occurs for $\sin\varphi = - H_{MX}/4[A_*(M, M) - A]$ which is a positive quantity, but φ is below $3°$. Since the nephelauxetic ratio is approximately $\beta = \cos^4\varphi = 0.994$ for $\varphi = 3°$, one would have to ascribe observed values close to $\beta = 0.98$ to modified (expanded) radial functions. Such a modification is capable of leaving ζ_{4f} almost constant. It is noteworthy that the configuration $b^2 a^1$ treated in Eqs. (17—20) contains only one $4f$-like electron, and hence, the dramatic decrease of φ is obtained without specific interelectronic repulsion in the partly filled shell.

However, one has to recognize a minor, but rather curious inaccuracy in the argument leading to Eq. (17). For instance, the configuration $b^2 a^2$ has the interelectronic repulsion

$$[A_*(M, M) + 4\, A_*(M, X) + A_*(X, X)]\, (1 - \sin^2\!\varphi \cos^2\!\varphi) \qquad (21)$$

oscillating between 75 and 100 percent of the correct value ($\varphi = 0°$) known to be invariant with φ (and hence without direct interest for us). It is necessary to consider the direct $J(a, b)$ and exchange $K(a, b)$ integrals of the two-electron operator ($6, 19, 114$) here being

$$\begin{aligned}
J(a, a) &= (\cos^4\!\varphi)\, J(M, M) + (2 \sin^2\!\varphi \cos^2\!\varphi)\, J(M, X) \\
&\quad + (4 \sin^2\!\varphi \cos^2\!\varphi)\, K(M, X) + (\sin^4\!\varphi)\, J(X, X) \\
J(a, b) &= (\sin^2\!\varphi \cos^2\!\varphi)\, J(M, M) + (\sin^4\!\varphi + \cos^4\!\varphi)\, J(M, X) \\
&\quad - (4 \sin^2\!\varphi \cos^2\!\varphi)\, K(M, X) + (\sin^2\!\varphi \cos^2\!\varphi)\, J(X, X) \qquad (22) \\
K(a, b) &= (\sin^2\!\varphi \cos^2\!\varphi\, J(M, M) - (2 \sin^2\!\varphi \cos^2\!\varphi)\, J(M, X) \\
&\quad + (\cos^4\!\varphi + \sin^4\!\varphi - 2 \sin^2\!\varphi \cos^2\!\varphi)\, K(M, X) + (\sin^2\!\varphi \cos^2\!\varphi)\, J(X, X) \\
J(b, b) &= (\sin^4\!\varphi)\, J(M, M) + (2 \sin^2\!\varphi \cos^2\!\varphi)\, J(M, X) \\
&\quad + (4 \sin^2\!\varphi \cos^2\!\varphi)\, K(M, X) + (\cos^4\!\varphi)\, J(X, X)
\end{aligned}$$

where $J(a, a)$ and $J(b, b)$ have the same coefficients as in the arguments involving A_* parameters, whereas $K(a, b)$ unexpectedly contains contributions from J integrals for $0° < \varphi < 90°$ and hence does not vanish even if the K integrals are negligible. *Harnung* and *Schäffer* (115) point out that $J(a, b) - K(a, b)$ for many purposes have simpler expressions than their parts, and this difference is here $J(M, X) - K(M, X)$. In the configuration $b^2\, a^2$, Eq. (22) gives the expression

$$\begin{aligned}
J(a, a) + 4\, J(a, b) &- 2\, K(a, b) + J(b, b) \\
&= J(M, M) + 4\, J(M, X) - 2\, K(M, X) + J(X, X)
\end{aligned} \qquad (23)$$

invariant with φ. It is noted that $A_*(M, X) = J(M, X) - \tfrac{1}{2} K(M, X)$. If the results in Eq. (22) are applied to the configuration $b^2\, a^1$ the two quantities $2 \sin^2\!\varphi$ to the left in Eq. (19) are halved with the consequence that the asymptotic Eq. (20) is replaced by

$$\sin\varphi = -(H_{MX}) / 2\,[A_*(M, M) - A + H_M - H_X] \qquad (24)$$

and the total energy of the configuration $b^2\, a^1$ is then stabilized ($\sin\varphi$) $H_{MX}/2$.

If the original arguments are applied to the configuration $b^4\, a^2$, the interelectronic repulsion between the six electrons is

$$\begin{aligned}
[1 + \sin^2\!\varphi(5 + \cos^2\!\varphi)]\, A_*(M, M) &+ [8 - 2 \sin^2\!\varphi \cos^2\!\varphi]\, A_*(M, X) \\
&+ [1 + \cos^2\!\varphi(5 + \sin^2\!\varphi)]\, A_*(X, X)
\end{aligned} \qquad (25)$$

and the expression analogous to Eq. (19) is

$$\begin{aligned}
(1 + 6 \sin^2\!\varphi)\, A_*(M, M) &+ (14 - 6 \sin^2\!\varphi)\, A_* + (2 + 2 \sin^2\!\varphi)\, H_M \\
&+ (4 - 2 \sin^2\!\varphi)\, H_X + (2\, \mathrm{tg}\varphi)\, H_{MX}
\end{aligned} \qquad (26)$$

For small φ, the differentiation with respect to $\sin\varphi$ situates the minimum energy close to

$$\sin\varphi = -\,(H_{\mathbf{MX}})\,/\,[6\,A_*(\mathrm{M,\,M}) - 6\,A + 2(H_{\mathbf{M}} - H_{\mathbf{X}})] \qquad (27)$$

which gives φ two-thirds as large as Eq. (20) for $H_{\mathbf{M}} = H_{\mathbf{X}}$. It is difficult to apply the new arguments in Eq. (22) to the configuration $b^4\,a^2$ before a closer analysis (6, 114) of $J(b,\,b)$ for the same and in two different b orbitals. If we restrict ourselves to replace

$$8\,A_*(a,\,b) + A_*(a,\,a) \quad \text{by} \quad 8\,J(a,\,b) - 4\,K(a,\,b) + J(a,\,a) \qquad (28)$$

the coefficient to $A_*(\mathrm{M,\,M})$ is $[1 + \sin^2\varphi(2 + 3\sin^2\varphi)]$ in Eq. (25) and $(1 + 2\sin^2\varphi)$ in Eq. (26) with the result that Eq. (24) is also obtained for $b^4\,a^2$.

It is striking that the denominators of Eqs. (20), (24) and (27) contain the difference $A_*(\mathrm{M,M}) - A$ rather analogous to the parameter U of *Mott* (85) and *Hubbard* (86) and quite foreign to *Hückel-Wolfsberg-Helmholz* models. It is perhaps also significant that the denominator is related to the first-order approximation (6, 39) to the electron transfer bands

$$h\nu_{\mathrm{e.\,t.}} = I(\mathrm{X}) - I(\mathrm{M}) + A_*(\mathrm{M,M}) - A_*(\mathrm{M,X}) \qquad (29)$$

where we have now learned (111, 112) that the difference $I(\mathrm{X}) - I(\mathrm{M})$ between the ionization energies may very well be negative, although $A_*(\mathrm{M,M})$ is highly positive.

Though the physical mechanism of the anti-ferromagnetic coupling in systems containing several transition-group ions described by equations such as Eq. (12) is very different from the weak covalent bonding in $4f$ group compounds described by Eq. (24), a common feature is that the denominator contains a large $A_*(\mathrm{M,M})$ with the result that the phenomenon is not primarily determined by one-electron energy differences. Though Eq. (24) undoubtedly is an approximation, it shows clearly that the minimization of the total energy in chromophores containing one or several $4f$ electrons has to take huge effects of interelectronic repulsion into account.

B. The Importance of Kinetic Energy

In the *Wolfsberg-Helmholz* model, the non-diagonal element $H_{\mathbf{MX}}$ to be multiplied by the overlap integral $S_{\mathbf{MX}}$ is assumed to be a constant k (between 1.5 and 2) times the arithmetic or geometric average $(-\,(H_{\mathbf{M}}\,H_{\mathbf{X}})^{1/2})$ of the diagonal elements $H_{\mathbf{M}}$ and $H_{\mathbf{X}}$ (100). Whereas the diagonal sum rule $(E_{\mathbf{a}} + E_{\mathbf{b}} = H_{\mathbf{M}} + H_{\mathbf{X}})$ would be valid for $k = 1$, the specific choice (6) $k = 2$ gives Eqs. (13) and (14) by second-order perturbation.

Hellmann and *Ruedenberg* (116) emphasized the importance of the kinetic operator for covalent bonding. In the asymptotic case of weak covalent bonding characterizing all $4f$ and some $3d$ group compounds, it is attractive to identify the non-diagonal H_{MX} with the local contribution to the kinetic energy in the bond region between the M and X atomic cores. Actually, the first paper (*117*) published in Chemical Physics Letters suggested to replace Eqs. (13) and (14) by

$$\psi_a = (\psi_M - \varkappa S_{MX}\ \psi_X)/[1 + (\varkappa^2 - 2\varkappa)S_{MX}^2]^{\frac{1}{2}}$$
$$\psi_b = (\psi_M + \lambda\ S_{MX}\ \psi_X)/[1 + (\lambda^2 + 2\lambda)S_{MX}^2]^{\frac{1}{2}} \tag{30}$$

where the delocalization parameters $\varkappa$ and λ are related by the normalization condition

$$\lambda = \varkappa - 1 + \varkappa\lambda S_{MX}^2 \tag{31}$$

In the limit of $\varkappa$ close to one, it is suggested (*19*) that the asymptotic *Slater* exponents μ_M and μ_X determine contributions to the kinetic energy:

$$E_a = H_M + \tfrac{1}{2}\ \varkappa\ S_{MX}^2(\mu_M^2 + \mu_X^2)$$
$$E_b = H_X - \tfrac{1}{2}\ \lambda\ S_{MX}^2(\mu_M^2 + \mu_X^2) \tag{32}$$

This approach has been further elaborated (*19*) and shown to yield the angular overlap model. It should not be concluded conversely that the validity of such results is a proof of asymptotic weak effects of the kinetic operator in Eq. (30). Thus, the order of the d-like orbitals in $3d$ group cyclopentadienides $M(C_5H_5)_2$ agrees with the perturbation of the equi-consequential contact potential at the ten carbon nuclei. Strictly speaking, this would suggest ten σ-bonds to the central M. However, there is no doubt that the "ligand field" effects in $4f$ group chromophores correspond to an additional nodal surface between M and the X in spite of the fact that the anti-bonding orbital ψ_a in Eq. (15) may have a higher $I(a)$ than the bonding, $I(b)$. If we conceive the minimization of energy of the chromophore by varying the extent of covalent bonding (as proposed by the LCAO model of heteronuclear systems) it is important to realize that $I(4f)$ may vary far more because of slightly changing fractional atomic charges than because of non-diagonal elements H_{MX} of the effective one-electron operator, and that it may be an overall advantage for the system that $I(4f)$ decreases to a certain extent, though not necessarily below the lowest I of the filled MO. It is not trivial that the numerous sub-levels of the various J-levels of $4f$ group compounds (*105*) can be consistently described by one-electron energy differences. Similar results have been obtained (*118*) for the $5f^1$ systems $PaCl_6^{-2}$, $PaBr_6^{-2}$, UF_6^-, UCl_6^-, UBr_6^- and NpF_6.

There is another aspect, where the kinetic energy operator is highly important in the $4f$ group. *Freeman* and *Watson* (*119*) performed *Hartree-Fock* calculations for various $4f$ group M^{+2} and M^{+3}. For an entirely different purpose (*120*) of comparing the relative intensities of photo-electron signals induced by 1486.6 eV photons, Dr. *Watson* was so kind as to indicate $\langle r^{-2}\rangle = 3.3$ bohr^{-2} for the half-filled $4f$ shell of Gd^{+3}. Since the angular part of the kinetic energy in spherical symmetry (*3, 19*) is $l(l+1)\langle r^{-2}\rangle/2$ atomic units, already the angular part is 540 eV in Gd^{+3} out of all proportion with $I_4 = 44.0$ eV of the gaseous ion or $I(4f)$ between 15.5 and 12 eV found in photo-electron spectra (see Table 2) of gadolinium

(III) compounds. Slightly less extreme behaviour is found in the $3d$ group (52) where $\langle r^{-2} \rangle$ is 2.13 bohr^{-2} in Fe^{+2} and 2.72 bohr^{-2} in Cu$^+$ giving the angular part of the kinetic energy 175 and 220 eV, respectively, about twenty times $I(3d)$ from photo-electron spectra. This large kinetic energy may very well be one of the facts determining the strict localization of $4f$ electrons in the elements and their compounds. If a change by 5 percent of $\langle r^{-2} \rangle$ modifies the kinetic energy twice $I(4f)$, it is another way of looking at the small φ from Eq. (15) in M(III) indicated by many physical measurements. Said in other words, the average radius of the partly filled $4f$ shell is far smaller than one would expect from its ionization energy.

C. Comparison with Copper(II) Complexes

It is interesting to compare the photo-electron spectra of $3d^9$ copper(II) and $4f^{13}$ ytterbium(III) which are chemically highly different, but have the *Mulliken electronegativity* (6) rather similar. It has been argued by many theorists that this average value of I and the electron affinity is *the* appropriate representation of one-electron quantities such as H_M and H_X. Since $I'(3d) = 11.1$ eV for copper(II) fluoride $(51, 112)$ and I' close to 9 eV for copper(I) compounds (such as CuCN) the *Mulliken* x is close to 10 eV. Correspondingly, $I'(4f)$ varying between 14 and 12 eV in Yb(III) and I' of Yb(II) expected (20) to be 6 eV suggests x somewhere between 10 and 9 eV. A much more extreme case is terbium(IV) where x is the mean value of 25 and 9 eV, close to 17 eV, neglecting a minor effect of the Franck-Condon principle.

Approximate LCAO calculations have been reported for the (hypothetical) planar CuF$_4^{-2}$ by *Tossell* and *Lipscomb* (121) and the planar CuCl$_4^2$ by *Demuynck, Veillard* and *Wahlgren* (122). In both cases, the twelve filled MO (constructed mainly from fluorine $2p$ or chlorine $3p$ orbitals) have Koopmans ε values some 11 to 3 eV less negative than the four filled $3d$-like orbitals. It is expected that the ionization energies show less pronounced differences, if the relaxation energy is taken into account (16). *Baerends* and *Ros* (123) discuss these deviations from *Koopmans* behaviour in iron(II) cyclopentadienide "ferrocene" Fe(C$_5$H$_5$)$_2$ where the three lowest I-values 6.9, 7.2 and 8.7 eV refer to two (degenerate) and one $3d$-like orbital, and to the two loosest bound MO mainly constituted of cyclopentadienide "π" orbitals. The *Koopmans* values (124) are 14.4, 16.6 and 11.7 eV, whereas the energy difference between gaseous Fe(C$_5$H$_5$)$_2^+$ and Fe(C$_5$H$_5$)$_2$ is calculated to be 8.3, 10.1 and 11.1 eV in the three cases. The Xα method $(98, 99)$ produces the more precise values (123) 6.7, 6.7 and 8.1 eV.

The comparison of Cu(II) and Yb(III) seems to show that the idea of *Mulliken* electronegativities is no universal remedy for the difficulties of the LCAO model. A less elegant and apparently somewhat arbitrary hypothesis is that one needs a convergent iteration, where the eigen-vectors in a highly covalent molecule are determined by the *Mulliken* x $= I($a$) - \frac{1}{2} A_*($a,a$)$ but in an almost electrovalent $4f$ group compound by I of the filled orbitals of the ligands and the electron affinity $I(4f) - A_*(4f,4f)$ of the partly filled $4f$ shell. Excepting final results known from chemical and spectroscopic intuition, such an iteration is not an easy problem, because the I values move quickly as a function of the inherently changing atomic charges. Though the concept of *differential ionization energies* (*3, 125*) cannot be directly applied (*6*) to the 4 *f* group compounds, it is imperative for such an iterative model to take the *Madelung* potential (*125*) into account, compensating the variation of I with varying fractional atomic charges, at least to a large extent.

Seen from this point of view, the negligible covalent bonding in $4f^6$ europium(III) is determined by the low $I(4f)$ of $4f^7$ europium(II) compounds for which $I' = 6.9$ eV in the colourless non-conductor $EuSO_4$ (*51*) and $I^* = 1.8$ eV corresponding to I close to 5 eV in the low-gap semiconductor EuS (*126*). These values between 5 and 7 eV are obviously far below I of most ligands in solid compounds. The similar argument does not hold for copper(II). It is quite characteristic (*112*) that the partly covalent bonding decreases the fractional atomic charges (*3, 6*) to such an extent that $I(3d)$ of $3d^6$ Co(III) is about a-third of $3d^{10}$ Ga(III), and of $3d^9$ Cu(II) some 5 eV below $3d^{10}$ Zn(II), whereas the moderate increase of $I(4f)$ from $4f^{13}$ Yb(III) to $4f^{14}$ Lu(III) agrees with the refined spin-pairing energy theory. The strongly anti-bonding character of the $(x^2 - y^2)$ like orbital containing one electron in copper(II) complexes producing rather specific stereochemical (*17, 19, 64*) and photo-electron (*51, 52*) characteristics is not at all comparable with the exceedingly weak anti-bonding character of some or all the seven $4f$ orbitals in a given chromophore (*105*) not disrupting the distribution of J, S, L-levels even in the configuration $4f^{q-1}$ studied in photo-electron spectra of lanthanide elements and compounds.

Note added in proof: Whereas thulium telluride (*79*) is essentially stoichiometric, *Campagna et al.* (*129*) recently studied the effect of gadolinium or thorium substitution in the semiconducting samarium(II) sulphide. The metallic NaCl-type $Gd_{0.14}Sm_{0.86}S$ and $Th_{0.15}Sm_{0.85}S$ show a photo-electron spectrum indicating comparable amounts of Sm[II] and Sm[III]. Other recent conclusions from photo-electron spectra have been reviewed (*130*).

Acknowledgments. The Swiss National Science Foundation provided the grant for the Varian-IEE 15, photo-electron spectrometer installed 1971. The writer is grateful for the careful experimental work by Dr. *Hervé Berthou* with the lanthanide compounds.

IV. References

1. *Moore, C. E.:* Atomic energy levels. Natl. Bur. Std. U.S. Circ. *467*, volumes 1,2,3. Washington, D. C.: 1949, 1952, 1958.
2. *Jørgensen, C. K.:* Gmelins Handbuch der anorganischen Chemie „Seltene Erden", Teil B, Lieferung 1.
3. *Jørgensen, C. K.:* Oxidation numbers and oxidation states. Berlin–Heidelberg–New York: Springer 1969.
4. *Humphreys, C. J., Meggers, W. F.:* J. Res. Natl. Bur. Std. *10*, 139 (1933).
5. *Rasmussen, E.:* Serier i de aedle luftarters spektre. Copenhagen: Danske Erhvervs Annoncebureau 1932.
6. *Jørgensen, C. K.:* Orbitals in atoms and molecules. London: Academic Press 1962.
7. *Condon, E. U., Shortley, G. H.:* Theory of atomic spectra (2. Ed.). Cambridge: University Press 1953.
8. *Sugar, J.:* J. Opt. Soc. Am. *53*, 831 (1963).
9. *Trees, R. E.:* J. Opt. Soc. Am. *54*, 651 (1964).
10. *Bryant, B. W.:* J. Opt. Soc. Am. *55*, 771 (1965).

11. *Sugar, J., Kaufman, V.:* J. Opt. Soc. Am. *62*, 562 (1972).
12. *Dieke, G. H., Crosswhite, H. M., Dunn, B.:* J. Opt. Soc. Am. *51*, 820 (1961).
13. *Dieke, G. H., Crosswhite, H. M.:* Appl. Opt. *2*, 675 (1963).
14. *Jørgensen, C. K.:* Angew. Chem. *85*, 1 (1973); Angew. Chem. Intern. Ed. Engl. *12*, 12 (1973).
15. *Racah, G.:* Phys. Rev. *76*, 1352 (1949).
16. *Jørgensen, C. K.:* Advan. Quantum Chem. *8*, 137 (1974).
17. *Jørgensen, C. K.:* Inorganic complexes. London: Academic Press 1963.
18. *Jørgensen, C. K.:* Struct. Bonding *1*, 234 (1966).
19. *Jørgensen, C. K.:* Modern aspects of ligand field theory. Amsterdam: North-Holland Publishing Co. 1971.
20. *Jørgensen, C. K.:* Struct. Bonding *13*, 199 (1973).
21. *Stoner, E. C.:* Phil. Mag. *48*, 719 (1924).
22. *Klinkenberg, P. F. A., Van Kleef, T. A. M.:* Physica *50*, 625 (1970).
23. *Brewer, L.:* J. Opt. Soc. Am. *61*, 1101 and 1666 (1971).
24. *Martin, W. C.:* J. Opt. Soc. Am. *61*, 1682 (1971).
25. *Martin, W. C.:* Optica Pura y Aplicada (Madrid) *5*, 181 (1972).
26. *Nugent, L. J., Vander Sluis, K. L.:* J. Opt. Soc. Am. *61*, 1112 (1971).
27. *Vander Sluis, K. L., Nugent, L. J.:* Phys. Rev. *A 6*, 86 (1972).
28. *Jørgensen, C. K.:* Lanthanides and elements from thorium to 184, in preparation.
29. *McClure, D. S., Kiss, Z.:* J. Chem. Phys. *39*, 3251 (1963).
30. *Drotning, W. D., Drickamer, H. G.:* J. Chem. Phys. *59*, 3482 (1973).
31. *Catalán, M. A., Röhrlich, F., Shenstone, A. G.:* Proc. Roy. Soc. (London) *A 221*, 421 (1954).
32. *Nugent, L. J., Baybarz, R. D., Burnett, J. L., Ryan, J. L.:* J. Inorg. Nucl. Chem. *33*, 2503 (1971).
33. *Nugent, L. J., Baybarz, R. D., Burnett, J. L., Ryan, J. L.:* J. Phys. Chem. *77*, 1528 (1973).
34. *Jørgensen, C. K.:* Mol. Phys. *5*, 271 (1962).
35. *Sugar, J., Reader, J.:* J. Chem. Phys. *59*, 2083 (1973).
36. *Loh, E.:* Phys. Rev. *147*, 332 (1966).
37. *Jørgensen, C. K.:* Mat. fys. Medd. Dan. Vidensk. Selskab *30*, no. 22 (1956).
38. *Stewart, D. C., Kato, D.:* Anal. Chem. *30*, 164 (1958).
39. *Jørgensen, C. K.:* Progr. Inorg. Chem. *12*, 101 (1970).
40. *Ryan, J. L., Jørgensen, C. K.:* J. Phys. Chem. *70*, 2845 (1966).
41. *Ryan, J. L.:* Inorg. Chem. *8*, 2053 (1969).
42. *Rosseinsky, D. R.:* Chem. Rev. *65*, 467 (1965).
43. *Jørgensen, C. K.:* Chimia (Aarau) *23*, 292 (1969).
44. *Berry, R. S., Reimann, C. W., Spokes, G. N.:* J. Chem. Phys. *37*, 2278 (1962).
45. *Jørgensen, C. K.:* Halogen Chem. *1*, 265, London: Academic Press 1967.
46. *Teegarden, K., Baldini, G.:* Phys. Rev. *155*, 896 (1967).
47. *Baldini, G.:* Phys. Rev. *128*, 1562 (1962).
48. *Roncin, J. Y., Damany, N., Vodar, B.:* Compt. Rend. (Paris) *260*, 96 (1965).
49. *Kratz, H. R.:* Phys. Rev. *75*, 1844 (1949).
50. *Beutler, H., Guggenheimer, K.:* Z. Physik *88*, 25 (1934).
51. *Jørgensen, C. K., Berthou, H.:* Mat. fys. Medd. Dan. Vidensk, Selskab *38*, no. 15 (1972).
52. *Jørgensen, C. K.:* Struct. Bonding *24*, 1 (1975).
53. *Jørgensen, C. K.:* Progr. Inorg. Chem. *4*, 73 (1962).
54. *Sugar, J.:* J. Opt. Soc. Am. *55*, 1058 (1965).
55. *Ballentine, L. E.:* Rev. Mod. Phys. *42*, 358 (1970).
56. *Ballentine, L. E.:* Foundations Phys. *3*, 229 (1973).
57. *Jørgensen, C. K.:* Theoret. Chim. Acta *34*, 189 (1974).
58. *Turner, D. W., Baker, C., Baker, A. D., Brundle, C. R.:* Molecular photoelectron spectroscopy. London: Interscience 1970.
59. *Siegbahn, K., Nordling, C., Johansson, G., Hedman, J., Hedén, P. F., Hamrin, K., Gelius, U., Bergmark, T., Werme, L. O., Manne, R., Baer, Y.:* ESCA applied to free molecules. Amsterdam: North-Holland Publishing Co. 1969.
60. *Manne, R., Åberg, T.:* Chem. Phys. Letters *7*, 282 (1970).
61. *Citrin, P. H., Hamann, D. R.:* Chem. Phys. Letters *22*, 301 (1973).

62. *Citrin, P. H., Hamann, D. R.:* Phys. Rev. *B 10,* 4948 (1974).

63. *Aarons, L. J., Guest, M. F., Hall, M. B., Hillier, I. M.:* Trans. Faraday Soc.(II) *69,* 563 (1973).

64. *Jørgensen, C. K.:* Topics Current Chem. *56,* 1 (1975).

65. *Wertheim, G. K., Rosencwaig, A., Cohen, R. L., Guggenheim, H. J.:* Phys. Rev. Letters *27,* 505 (1971).

66. *Cox, P. A., Baer, Y., Jørgensen, C. K.:* Chem. Phys. Letters *22,* 433 (1973).

67. *Cox, P. A.:* Struct. Bonding *24,* 59 (1975).

68. *Hall, H. T., Merril, L.:* Inorg. Chem. *2,* 618 (1963).

69. *Vergand, F., Bonnelle, C.:* Solid State Commun. *10,* 397 (1972).

70. *Hedén, P. O., Löfgren, H., Hagström, S. B. M.:* Phys. Rev. Letters *26,* 432 (1971).

71. *Hagström, S. B. M., Brodén, G., Hedén, P. O., Löfgren, H.:* J. Phys. (Paris) *C 4,* 269 (1971).

72. *Baer, Y., Busch, G.:* J. Electron Spectr. (Namur Conference 1974) *5,* 611 (1974).

73. *Carnall, W. T., Fields, P. R., Rajnak, K.:* J. Chem. Phys. *49,* 4412, 4424, 4443, 4447 and 4450 (1968).

74. *Crosswhite, H. M., Schwiesow, R. L., Carnall, W. T.:* J. Chem. Phys. *50,* 5032 (1950).

75. *Cox, P. A., Evans, S., Orchard, A. F.:* Chem. Phys. Letters *13,* 386 (1972).

76. *Herbst, J. F., Lowy, D. N., Watson, R. E.:* Phys. Rev. *B 6,* 1913 (1972).

77. *Hüfner, S., Wertheim, G. K.:* Phys. Rev. *B 7,* 5086 (1973).

78. *Campagna, M., Bucher, E., Wertheim, G. K., Buchanan, D. N. E., Longinotti, L. D.:* Proceed. 11. Rare Earth Conference (Traverse City, Michigan, 1974) p. 810.

79. *Campagna, M., Bucher, E., Wertheim, G. K., Buchanan, D. N. E., Longinotti, L. D.:* Phys. Rev. Letters *32,* 885 (1974).

80. *Jørgensen, C. K.:* Mol. Phys. *7,* 417 (1964).

81. *Hulliger, F.:* Helv. Phys. Acta *41,* 945 (1968).

82. *Jørgensen, C. K.:* Acta Chem. Scand. *9,* 540 (1955).

83. *Löwdin, P. O.:* J. Phys. Chem. *61,* 55 (1957).

84. *Jørgensen, C. K.:* Solid State Phys. *13,* 375 (1962).

85. *Mott, N. F.:* Proc. Phys. Soc. (London) *A 62,* 416 (1949).

86. *Hubbard, J.:* Proc. Roy. Soc. (London) *A 277,* 237 (1964).

87. *Davidson, E. R., Jones, L. L.:* J. Chem. Phys. *37,* 2966 (1962).

88. *Moffitt, W.:* Proc. Roy. Soc. (London) *A 210,* 224 and 245 (1951).

89. *Griffith, J. S.:* Struct. Bonding *10,* 87 (1972).

90. *Glerup, J.:* Acta Chem. Scand. *26,* 3775 (1972).

91. *Sugano, S., Tanabe, Y., Kamimura, H.:* Multiplets of transition-metal ions in crystals. New York: Academic Press 1970.

92. *Veal, J. T., Jeter, D. Y., Hempel, J. C., Eckberg, R. P., Hatfield, W. E., Hodgson, D. J.:* Inorg. Chem. *12,* 2928 (1973).

93. *Iwashita, T., Idogaki, T., Uryû, N.:* J. Phys. Soc. Japan *30,* 1587 (1971).

94. *Bang, E.:* Acta Chem. Scand. *22,* 2671 (1968).

95. *Lidiard, A. B.:* Rept. Progr. Phys. *17,* 201 (1954).

96. *Rado, G. T., Suhl, H.:* Magnetism. New York: Academic Press 1963—65.

97. *Schuchert, H., Hüfner, S., Faulhauber, R.:* J. Appl. Phys. *39,* 1137 (1968).

98. *Slater, J. C.:* Advan. Quantum Chem. *6,* 1 (1972).

99. *Johnson, K. H.:* Advan. Quantum Chem. 7, 143 (1973).

100. *Basch, H., Viste, A., Gray, H. B.:* J. Chem. Phys. *44,* 10 (1966).

101. *Cotton, F. A., Harris, C. B.:* Inorg. Chem. *6,* 369 and 376 (1967).

102. *Cox, P. A., Evans, S., Hamnett, A., Orchard, A. F.:* Chem. Phys. Letters *7,* 414 (1970).

103. *Evans, S., Hamnett, A., Orchard, A. F., Lloyd, D. R.:* Discussions Faraday Soc. *54,* 227 (1973).

104. *Jørgensen, C. K.:* Chem. Phys. Letters *2,* 549 (1968).

105. *Jørgensen, C. K., Pappalardo, R., Schmidtke, H. H.:* J. Chem. Phys. *39,* 1422 (1963).

106. *Schäffer, C. E., Jørgensen, C. K.:* Mol. Phys. *9,* 401 (1965).

107. *Schäffer, C. E.:* Struct. Bonding *5,* 68 (1968).

108. *Schäffer, C. E.:* Pure Appl. Chem. *24,* 361 (1970).

109. *Schäffer, C. E.:* Struct. Bonding *14,* 69 (1973).

110. *Schäffer, C. E.:* Theoret. Chim. Acta *34,* 237 (1974).

111. *Jørgensen, C. K.:* Chimia (Aarau) *27*, 203 (1973).
112. *Jørgensen, C. K.:* Chimia (Aarau) *28*, 6 (1974).
113. *Burns, G.:* Phys. Letters *25 A* 15 (1967).
114. *Jørgensen, C. K.:* Acta Chem. Scand. *12*, 903 (1958).
115. *Harnung, S. E., Schäffer, C. E.:* Struct. Bonding *12*, 201 and 257 (1972).
116. *Ruedenberg, K.:* Rev. Mod. Phys. *34*, 326 (1962).
117. *Jørgensen, C. K.:* Chem. Phys. Letters *1*, 11 (1967).
118. *Edelstein, N., Brown, D., Whittaker, B.:* Inorg. Chem. *13*, 563 (1974).
119. *Freeman, A. J., Watson, R. E.:* Phys. Rev. *127*, 2058 (1962).
120. *Jørgensen, C. K., Berthou, H.:* Discussions Faraday Soc. *54*, 269 (1973).
121. *Tossell, J. A., Lipscomb, W. N.:* J. Am. Chem. Soc. *94*, 1505 (1972).
122. *Demuynck, J., Veillard, A., Wahlgren, U.:* J. Am. Chem. Soc. *95*, 5563 (1973).
123. *Baerends, E. J., Ros, P.:* Chem. Phys. Letters *23*, 391 (1973).
124. *Coutière, M. M., Demuynck, J., Veillard, A.:* Theoret. Chim. Acta *27*, 281 (1972).
125. *Jørgensen, C. K., Horner, S. M., Hatfield, W. E., Tyree, S. Y.:* Intern. J. Quantum Chem. *1*, 191 (1967).
126. *Eastman, D. E., Kuznietz, M.:* J. Appl. Phys. *42*, 1396 (1971).
127. *Sugar, J., Kaufman, V.:* J. Opt. Soc. Am. *64*, 1656 (1974).
128. *Litzén, U.:* Physica Scripta (Stockholm) *10*, 103 (1974).
129. *Campagna, M., Bucher, E., Wertheim, G. K., Longinotti, L. D.:* Phys. Rev. Letters *33*, 165 (1974).
130. *Jørgensen, C. K.:* Chimia (Aarau) *29*, 53 (1974).

Received September 24, 1974

The Intensities of Lanthanide $f \leftrightarrow f$ Transitions

Robert D. Peacock

Department of Chemistry, King's College, University of London, Strand, London WC2R 2LS, England

Table of Contents

1. Introduction

Interest in the intensities of lanthanide $f \leftrightarrow f$ spectra can be said to have begun with a paper published in 1937 by *Van Vleck* (*1*). At that time it had not been established whether the sharp lines in the spectra of trivalent lanthanide ions were in fact due to transitions within the $4f^N$ configuration or due to transitions between that configuration and one of higher energy (for example $4f^{N-1} 5d$). In the latter case the transitions would occur by an allowed electric dipole mechanism; in the former by a forced electric dipole (involving either a static or a vibronic perturbation) or a magnetic dipole or electric quadrupole mechanism. *Van Vleck* calculated (*1*) the intensity expected on the basis of each of these possibilities and concluded that, whilst that produced by the allowed electric dipole mechanism was much too large (ruling out $f \rightarrow d$ transitions), any of the other mechanisms could account for the observed intensities. He interpreted the many 'extra' lines found in the spectra of most solid lanthanide compounds as being due to vibronic transitions and this gave weight to his conclusion that much of the intensity was due to the vibronically induced absorption of electric dipole radiation. He also noted that since the $f \leftrightarrow f$ transitions of Eu(III), Gd(III) and Tb(III) are spin forbidden in zero order, magnetic dipole transitions should be relatively of more importance in the spectra of these ions.

The next step forward took place in 1942 when the group of spectroscopists at the Zeeman Laboratories, University of Amsterdam, began a systematic study of the solution spectra of the lanthanide aquo ions (*2*). As well as giving the first accurate measurements of the absolute intensities of lanthanide spectra (helped by the use of photoelectric instead of photographic techniques) an important paper (*3*) was published, on the origin of the intensities, which reached somewhat different conclusions from those of *Van Vleck* (*1*). In particular they calculated (*3*) that electric quadrupole transitions would have much less intensity than is observed, and that magnetic dipole transitions would only be observed if the transition could not obtain intensity by a forced electric dipole mechanism. To exemplify the latter conclusion they instanced the now well-known examples of the $^5D_1 \leftarrow {}^7F_0$ and $^5D_0 \rightarrow {}^7F_1$ transitions of Eu(III). They further concluded that the intensity due to vibronically induced electric dipole transitions should be comparable to, but probably somewhat less than, the corresponding intensity induced by a static perturbation. They then calculated the oscillator strengths of the Pr^{3+}, Tm^{3+} and Yb^{3+} aquo ions considering only a static forced electric dipole mechanism and concluded that except for two transitions, one in the spectrum of Pr^{3+} assigned as $^1I_6 \leftarrow {}^3H_4$ and one in that of Tm^{3+} assigned as $^1I_6 \leftarrow {}^3H_6$, this mechanism could account for all the intensity. In fact the two 'anomalies' can be explained: in the Pr^{3+} spectrum the anomalous transition should be assigned to $^1D_2 \leftarrow {}^3H_4$ and in the case of Tm^{3+} it appears that an impurity absorption had been measured. The conclusions of this paper, although based on semiquantitative calculations, are broadly in agreement with present day opinion.

A compilation of the oscillator strengths of all the aquo ions appeared in 1948 (*4*) and apart from an unpublished report (*5*) in 1952 no other intensity data appear in the literature until after the publication of the theory of lanthanide intensities by *Judd* (*6*) and *Ofelt* (*7*) in 1962.

2. Forced Electric Dipole Transitions — The Judd-Ofelt Theory

2.1. Introduction

All recent studies of the intensities of $f \leftrightarrow f$ spectra have been prompted by this theory, published independently in 1962 by *Judd* (6) and *Ofelt* (7) and now known as the Judd-Ofelt theory. Since virtually every paper published since then on the subject of lanthanide intensities has been concerned with either testing the theory or using it to make structural predictions or electronic assignments it is felt that it must be discussed in some detail. In particular it is increasingly common to find authors uncritically using Judd's final equation as the starting point of their discussion. The theory is necessarily couched in the formalism of tensor operators, $n-j$ symbols and reduced matrix elements and since many chemists may be unfamiliar with these it is proposed to give a brief account of the major definitions and relationships. The reader is referred to the original papers of *Racah* (8) and to the more readable books by *Edmonds* (9), *Judd* (10) and *Wybourne* (11). A particularly good account of these subjects and their application to atomic spectroscopy is to be found in *Shore* and *Menzel* (12).

2.2. Tensor Operators

An irreducible tensor operator of rank k, $T^{(k)}$ is defined as that collection of components

$$T_k^{(k)}, T_{k-1}^{(k)} \ldots \ldots, T_q^{(k)}, \ldots \ldots T_{-k}^{(k)}$$

which transform under rotations according to the equation

$$T_q^{(k)}(\text{new}) = \sum_{q'} \mathscr{D}_{qq'}^{(k)} T_{q'}^{(k)} (\text{old})$$

where $D^{(k)}$ is a general rotation matrix of rank k.

In particular *Racah* defined (8) irreducible tensors, $C^{(k)}$, which transform as spherical harmonics, having components

$$C_q^{(k)} = \left(\frac{4\pi}{2k+1} \right)^{\frac{1}{2}} Y_{kq}$$

where Y_{kq} is a spherical harmonic of rank k.

$C^{(0)}$, then, has only one component, $C_0^{(0)} = (4\pi)^{\frac{1}{2}} Y_{00} = 1$ and is a scalar. The position vector $\boldsymbol{r}$ is a tensor of rank 1 and so may be related to $C^{(1)}$ thus:

$$\boldsymbol{r} = r\, C^{(1)}$$

having components

$$r_{+1} = r\, C_1^{(1)} = r\, (4\pi/3)^{\frac{1}{2}} Y_{11} = {}^{-1}/\sqrt{2}\, (x + iy)$$
$$r_0 = r\, C_0^{(1)} = r\, (4\pi/3)^{\frac{1}{2}} Y_{10} = z$$
$$r_{-1} = r\, C_{-1}^{(1)} = r\, (4\pi/3)^{\frac{1}{2}} Y_{1-1} = {}^{1}/\sqrt{2}(x - iy)$$

The electric dipole moment operator $\boldsymbol{P}$ is the sum over all electrons of the position vectors of these electrons

$$\boldsymbol{P} = -e\sum_i \boldsymbol{r}_i = -e\sum_i r_i(\boldsymbol{C}^{(1)})_i$$

and is written $-e\boldsymbol{D}^{(1)}$ with components $-e\boldsymbol{D}_q^{(1)}$ where $q=0, \pm 1$

Another operator we will encounter later is the crystal field operator $\boldsymbol{V}^{\text{C.F.}}$. This in general is of rank t.

$$\boldsymbol{V}^{\text{C.F.}} = \sum_{t,p} A_{tp} \sum_i r_i(\boldsymbol{C}_p^{(t)})_i$$

It is written $\sum_{t,p} A_{tp} \boldsymbol{D}_p^{(t)}$; A_{tp} are called the crystal field parameters.

2.3. Matrix Elements and Reduced Matrix Elements

We will be interested in evaluating matrix elements of the type

$$\langle \alpha SLJM | T_q^{(k)} | \alpha'S'L'J'M' \rangle$$

where α represents all quantum numbers other than S,L,J and M which are required to completely specify the state and $M = M_J$. It can be seen that since there are $2k+1$ components of the operator and $2J+1$ values of M, there will be in general $(2k+1)(2J+1)(2J'+1)$ separate matrix elements to evaluate for every $J \leftrightarrow J'$ transition. Consider for example the matrix element

$$\langle {}^5I_8 \, M \, | T_q^{(2)} | {}^5H_6 \, M' \rangle \text{ of } H_0^{3+};$$

there are 1105 separate matrix elements to be evaluated.

In order to greatly simplify this problem we can apply the Wigner-Eckart theorem and write the matrix element as the product of a *reduced* matrix element which depends upon J, J' and k but not upon M,M' and q, *i.e.* is component independent, and a 3—j symbol whose value depends upon the components considered.

$$\langle \alpha SLJM | T_q^{(k)} | \alpha'S'L'J'M' \rangle = (-1)^{J-M} \begin{pmatrix} J & k & J' \\ -M & q & M' \end{pmatrix} \langle \alpha SLJ \| T^{(k)} \| \alpha'S'L'J' \rangle \quad (1)$$

3—j symbols, which are written in general as

$$\begin{pmatrix} j_1 & j_2 & j_3 \\ m_1 & m_2 & m_3 \end{pmatrix}$$

have useful symmetry properties, and (in particular) are zero unless $m_1 + m_2 + m_3 = 0$ and $j_1 + j_3 \geqslant j_2 \geqslant |j_1 - j_3|$. The values of all necessary 3—j symbols have been tabulated (13). The reduced matrix elements may be further reduced by an analogous process involving a 6—j symbol.

$$\langle \alpha S L J \| T^{(k)} \| \alpha' S' L' J' \rangle = \delta(SS') (-1)^{S+L'+J+k} [(2J+1)(2J'+1)]^{\frac{1}{2}}$$

$$\times \begin{Bmatrix} L & k & L' \\ J' & S & J \end{Bmatrix} \langle \alpha S L \| T^{(k)} \| \alpha' S' L' \rangle \qquad (2)$$

The values of the $6j$ symbols have been tabulated (13) and those of the doubly reduced matrix elements have been calculated and published (14) for the $4f^N$ configuration.

2.4. The Judd-Ofelt Theory (6, 7)

The oscillator strength, P, of a component of an electric dipole transition from a ground state $|A\rangle$ to an excited state $|B\rangle$ is given by the equation:

$$P = \chi \left[\frac{8 \pi^2 m c \sigma}{h} \right] |\langle A | D_q^{(1)} | B \rangle|^2 \qquad (3)$$

where m is the electron mass, h Plank's constant, c the velocity of light, σ the energy of the transition in cm^{-1} and χ is the Lorentz field correction for the refractivity of the medium. $D_q^{(1)}$ is defined in Section 2.2. above. The matrix elements of the electric dipole operator vanish between states of the same parity and so between states arising from the same configuration.

In the free ion approximation the states of the $4f^N$ configuration are taken as linear combinations of Russell-Saunders coupled states $|f^N \alpha S L J\rangle$:

$$|f^N \alpha [SL] J \rangle = \sum_{S,L} A(S,L) |f^N \alpha S L J \rangle \qquad (4)$$

In these 'intermediate coupled' states the quantum numbers S and L, although convenient for purposes of labelling, are not 'good' and are enclosed in square brackets. To save space the wavefunction defined in Eq. (4) will frequently be written $|f^N \psi J\rangle$.

The matrix of the electric dipole operator, then, vanishes between these states. In order to 'force' an electric dipole transition it is necessary to mix into the $4f^N$ configuration another configuration having opposite parity. Such mixing may be accomplished by the odd parity terms of the expansion of the crystal field potential.

$$V^{\text{C.F.}} = \sum_{t,p} A_{tp} D_p^{(t)} , \text{ with } t \text{ odd.}$$

Considering the crystal field as a first order perturbation and mixing in states of a higher energy opposite parity configuration $|nl\alpha''[S''L'']J''M''\rangle$ (which will be written $|\psi''\rangle$) we may write $|A\rangle$ and $|B\rangle$ as follows:

$$|A\rangle = |f^N \psi J M\rangle + \sum_k \frac{\langle \psi'' | \langle f^N \psi J M | V^{\text{C.F.}} | \psi'' \rangle}{E(4f^N J) - E(\psi'')}$$

87

$$|B\rangle = |f^N \psi' J' M'\rangle + \sum_k \frac{\langle \psi'' | \langle f^N \psi' J' M' | V^{C. F.} | \psi''\rangle}{E(4f^N J') - E(\psi'')}$$

where k stands for all the quantium number of the excited configuration. The dipole strength $D = e^2 \langle A | D_q^{(1)} | B \rangle^2$ of a transition from $|A\rangle$ to $|B\rangle$ is then:

$$D = \left[e \sum_{k,t,p} A_{tp} \left\{ \frac{\langle f^N \psi J M | D_q^{(1)} | \psi'' \rangle \langle \psi'' | D_p^{(t)} | f^N \psi' J' M' \rangle}{E(4f^N J') - E(\psi'')} \right. \right.$$
$$\left. \left. + \frac{\langle f^N \psi J M | D_p^{(t)} | \psi'' \rangle \langle \psi'' | D_q^{(1)} | f^N \psi' J' M' \rangle}{E(4f^N J) - E(\psi'')} \right\} \right]^2 \qquad (5)$$

The problem now is to reduce this perturbation expression to usable form We will consider initially only the first half of Eq. (5). The treatment of the second half is analogous.

Firstly we can express the operators $D_q^{(1)}$ and $D_p^{(t)}$ as $\sum_i r_i (C_q^{(1)})_i$ and $\sum_i r_i^t (C_p^{(t)})_i$ respectively and so take out radial integrals. The expression then becomes:

$$e \sum_{k,t,p} A_{tp} \langle f^N \psi J M | \sum_i (C_q^{(1)})_i | \psi'' \rangle \langle \psi'' | \sum_i (C_p^{(t)})_i | f^N \psi' J' M' \rangle \langle 4f | r | nl \rangle \langle nl | r^t | 4f \rangle$$
$$\times [E(4f^N J') - E(\psi'')]^{-1} \qquad (6)$$

where $\langle nl | r^k | n' l' \rangle$ is an abbreviation for

$$\int_0^\infty \mathscr{R}(nl) \, r^k \, \mathscr{R}(n' l') dr$$

and $\mathscr{R}/r$ is the radial part of the appropriate one electron wave function.

One way of simplifying this expression is to use the procedure known as closure (15). If the energy of the perturbing configuration is invarient with respect to α'', S'', L'', J'' and M'' the equation:

$$\sum_{\alpha'' S'' L'' J'' M''} \langle f^N \psi J M | \sum_i (C_q^{(1)})_i | \psi'' \rangle \langle \psi'' | \sum_i (C_p^{(t)})_i | f^N \psi' J' M' \rangle$$
$$= (-1)^{p+q+\lambda} [\lambda] \begin{pmatrix} 1 & \lambda & t \\ q & -p & -q \, p \end{pmatrix} \times \langle f^N \psi J M | \sum_i ((C_q^{(1)} C_p^{(t)})_{-p-q}^{(\lambda)})_i | f^N \psi' J' M' \rangle \qquad (7)$$

would be exact. The abbreviation $[k] = (2k + 1)$ has been introduced. The combined operator

$$(C_q^{(1)} C_p^{(t)})_{-p-q}^{(\lambda)}$$

may then be further simplified and Eq. (7) becomes

$$(-1)^{p+q} (-1)^{f+l} [\lambda] [f] [l] \begin{pmatrix} 1 & \lambda & t \\ q & -p & -q \, p \end{pmatrix} \begin{Bmatrix} 1 & t & \lambda \\ f & f & l \end{Bmatrix} \langle f \| C^{(1)} \| l \rangle \langle l \| C^{(t)} \| f \rangle$$
$$\times \langle f^N \psi J M | U_{-p-q}^{(\lambda)} | f^N \psi' J' M' \rangle \qquad (8)$$

$U^{(\lambda)} = \sum u_i^{(\lambda)}$, where $u_i^{(\lambda)}$ is defined by $\langle nl \| u_i^{(\lambda)} \| n'l' \rangle = \delta(nn')\delta(ll')$ and so operates within a configuration. *Judd* made the assumption that the energies of the excited configurations are independent of all quantum numbers except n and l; that is that the excited configurations are (independently) completely degenerrate. This is an admitted weak link in the theory. The final simplification is, however, probably the least justified. This involves equating $E(4f^N J) - E(\psi'')$ and $E(4f^N J') - E(\psi'')$ and replacing them by an average energy denominator $\Delta E(\psi'')$. This is clearly a rather bad approximation for certain of the lanthanides where the energy of, say, the $4f^{N-1} 5d$ configuration is not very much greater than that of the $J \leftrightarrow J'$ transitions being considered. However it does result in significant simplification. If the procedures outlined above [Eqs. (6, 7) and (8)] are applied to both halves of Eq. (5) the result for each half will differ only in the 3-j symbol. Because of the symmetry relation (6)

$$\begin{pmatrix} 1 & \lambda & t \\ q & -p & -q\,p \end{pmatrix} = (-1)^{1+\lambda+t} \begin{pmatrix} t & \lambda & 1 \\ p & -p & -q\,q \end{pmatrix}$$

the two halves are equal if λ is even and cancel exactly if λ is odd. Thus by introducing the approximation of an average energy denominator we can remove all the terms involving odd λ. The 6-j symbol restricts λ to values of less than or equal to 6. Finally, then, we obtain

$$D = \left[e \sum_{p,t,\text{even}\lambda} (-1)^{p+q} A_{tp}[\lambda] \Xi(t,\lambda) \begin{pmatrix} 1 & \lambda & t \\ q & -p & -q\,p \end{pmatrix} \langle f^N \psi J M | U_{-p-q}^{(\lambda)} | f^N \psi' J' M' \rangle \right]^2$$

(9)

where

$$\Xi(t,\lambda) = 2 \sum (-1)^{f+l}[f][l] \begin{Bmatrix} 1 & \lambda & t \\ f & l & f \end{Bmatrix} \langle f \| C^{(1)} \| l \rangle \langle l \| C^{(t)} \| f \rangle \langle 4f | r | nl \rangle$$
$$\langle nl | r^t | 4f \rangle \Delta E(\psi'')^{-1}$$

(10)

and there is an implicit summation over n and l of all configurations which it is desired to mix in. The matrix element in Eq. (9) can now be reduced [via Eq. (1)] and the resulting expression for D inserted in Eq. (3) to give

$$P_{\text{E.D.}} = \chi \left[\frac{8\pi^2 m c \sigma}{h} \right] \left[\sum_{p,t\,\text{even}\,\lambda} (-1)^{p+q}[\lambda] A_{tp} \begin{pmatrix} 1 & \lambda & t \\ q & -p & -q\,p \end{pmatrix} \begin{pmatrix} J & \lambda & J' \\ -M & -q & -p\,M' \end{pmatrix} \Xi(t,\lambda) \right.$$
$$\left. \langle f^N \psi J \| U^{(\lambda)} \| f^N \psi' J' \rangle \right]^2$$

(11)

In this form (equation 11) the theory is equipped to deal with transitions between individual Stark levels (whose wavefunctions are in general linear combinations of $|f^N \alpha[SL]JM\rangle$ states).

In solution, however, such transitions cannot usually be distinguished and it is convenient to sum over all the Stark levels of the ground state (assuming all are equally populated). At the same time we can sum over the components of $D_q^{(1)}$ and the components of $D_p^{(t)}$ which is appropriate for isotropic light. The 3-j symbols in Eq. (11) vanish and are thus replaced by a factor of $3^{-1}(2J+1)^{-1}$ $(2t+1)^{-1}$. Finally, then, we have,

$$P_{\text{E.D.}} = \sum_{\lambda=2,4,6} \sigma \mathscr{T}_\lambda \langle f^N \alpha[SL]J \| U^{(\lambda)} \| f^N \alpha'[S'L']J' \rangle^2 (2J+1)^{-1} \quad (12\text{a})$$

where

$$\mathscr{T}_\lambda = \chi \left[\frac{8\pi^2 m c}{3h} \right] [\lambda] \sum_{p,t} |A_{tp}|^2 \, \Xi^2(t,\lambda)(2t+1)^{-1} \quad (12\text{b})$$

This notation *(16, 17)* has generally been adopted for solution studies and the Judd-Ofelt parameters, $\mathscr{T}_\lambda$, are slightly different from those used in Judd's original paper, T_λ. They are related by

$$T_\lambda = \mathscr{T}_\lambda (2J+1)^{-1} c^{-1}$$

For work involving the isotropic spectra of ions in crystals, many workers have used an alternative notation. Because the refractive index correction χ varies with wavelength it is not incorporated in the Judd-Ofelt parameters but left explicitly outside the summation sign. In this case

$$P_{\text{E.D.}} = \chi \left[\frac{8\pi^2 m c}{3h} \right] \sigma \sum_{\lambda=2,4,6} \Omega_\lambda \langle f^N \alpha[SL]J \| U^{(\lambda)} \| f^N \alpha'[S'L']J' \rangle^2 (2J+1)^{-1} \quad (13\text{a})$$

where

$$\Omega_\lambda = [\lambda] \sum_{p,t} |A_{tp}|^2 \, \Xi^2(t,\lambda)(2t+1)^{-1} \quad (13\text{b})$$

In this review Ω_λ notation will be used since it is always possible to convert $\mathscr{T}_\lambda$ into Ω_λ but not vice versa. The units of T_λ are sec^{-1}, of $\mathscr{T}_\lambda$, cm and of Ω_λ, cm^2.

The analogous quantity to the oscillator strength normally used in dealing with emission spectra is the spontaneous emission coefficient, A.

$$A_{\text{E.D.}} = \chi \left[\frac{64\pi^4 e^2}{3h} \right] \sigma^3 \sum_{\lambda=2,4,6} \Omega_\lambda \langle f_\alpha^N[SL]J \| U^{(\lambda)} \| f_{\alpha'}^N[S'L']J' \rangle^2 (2J+1)^{-1} \quad (14)$$

and Ω_λ is defined in Eq. (13b). No distinction will be made between Ω_λ measured by means of absorption or emission spectra. The normal spectroscopic convention will be used when discussing individual transitions: the arrow connecting states goes from right to left for absorption and from left to right for emission. The reduced matrix elements in Eq. (12a) and (13a) will frequently be written $\Gamma^{(\lambda)}$.

2.5. Vibronic Interactions

As well as the static crystal field potential, there exists another possible mechanism whereby configurations of opposite parity may be mixed into the $4f^N$ configuration: that is the potential produced by non-totally symmetric vibrations.

If the normal co-ordinates of the vibrating complex (lanthanide ion plus ligands) are denoted Q_j and the totality of vibrational quantum numbers by η (ground) and η' (excited state), then the potential $V^{\text{VIB.}}$ may be written:

$$V^{\text{VIB}} = \sum_{t,\,p} \sum_j \frac{\partial A_{tp}}{\partial Q_j} Q_j \, \boldsymbol{D}_p^{(t)}$$

Writing the perturbed wavefunctions $|A\rangle$ and $|B\rangle$ of Section 2.4 as $|A\,\eta\rangle$ and $|B\,\eta'\rangle$ we have:

$$\langle A\,\eta\,|\boldsymbol{D}_q^{(1)}|B\,\eta'\rangle = \sum_j \langle \eta\,|Q_j|\,\eta'\rangle \frac{\partial}{\partial Q_j} \langle A\,|\boldsymbol{D}_q^{(1)}|B\rangle$$

Carrying the analysis through exactly as before and writing the probability that the complex is in the state η by $\varrho(\eta)$, the vibronic oscillator strength for isotropic solution may be written

$$P_{\text{VIB.}} = \chi \left[\frac{8\,\pi^2\,m\,c}{3\,h}\right] \sigma \sum_{\lambda=2,4,6} \Omega_\lambda(\text{VIB.})\,(\Gamma^{(\lambda)})^2\,(2J+1)^{-1} \qquad (15\text{a})$$

where

$$\Omega_\lambda(\text{VIB.}) = [\lambda] \sum_{p,t,j,\eta,\eta'} \left|\frac{\partial A_{tp}}{\partial Q_j}\right|^2 \Xi^2\,(t,\lambda)\,\langle \eta\,|Q_j|\,\eta'\rangle^2\,\varrho(\eta)\,(2t+1)^{-1} \qquad (15\text{b})$$

which is the vibronic analogue of Eq. (13).

The vibronic contribution to the electric dipole oscillator strength has exactly the same form as the static contribution and the Ω_λ parameters as obtained experimentally will contain contributions from both mechanisms.

2.6. Assumptions Made in the Derivation of the Theory

It is worth mentioning once again the three assumptions made in the derivation of Eqs. (12) and (13).

1. The configurations which are mixed in to the $4f^N$ configuration ($4f^{N-1}5d$ and $4f^{N-1}ng$ are the lowest energy odd-parity ones) have been assumed to be degenerate. The positions and energies of the low-lying configurations of the lanthanides are known, and transitions to the $4f^{N-1}5d$ configuration, for example, start at around 60,000 cm^{-1} for most of the lanthanides(III) and are still observed at 100,000 cm^{-1}, the present limit of measurement. Thus the states of the $4f^{N-1}5d$ configuration are spread over at least 40,000 cm^{-1}. It has been argued (6) that the extreme complexity of the configuration should make the approximation more valid than it would seem. The breakdown of this validity has been invoked to explain the unusual spectral behaviour of Pr(III). This is discussed separately in Section 6 below.

2. The energy differences $\Delta_1 = E(f^N J) - E(\psi'')$ and $\Delta_2 = E(f^N J') - E(\psi'')$ have been assumed equal. This approximation enabled us to cancel out the parts of the intensity expression having λ odd. It is roughly valid for those lanthanides where the excited configurations are at quite high energies but much less so for, say, Pr(III) where the $4f^{N-1}5d$ configuration starts at about 45,000 cm^{-1}. Consider $E(f^N J) = 0$, $E(f^N J') = 25,000$ cm^{-1} and $E(\psi'') = 80,000$ cm^{-1} then the ratio $(\Delta_1 + \Delta_2)^{-1} : (\Delta_1 - \Delta_2)^{-1}$ is $\sim 6:1$. For $E(\psi'') = 45.000$ cm^{-1}, however, the ratio is $\sim 3.5:1$ and so in this case considerable intensity may come from terms having λ odd.

3. All the M_J levels of the ground state have been assumed equally populated. The splitting of the ground state of some complexes is as high as 600 cm^{-1} and so even at room temperature there will be considerable inequality of population of the Stark components. Because each Stark component is a linear combination of M_J states, however, the approximation may be quite good even for the large ground state splittings mentioned. The Judd-Ofelt theory cannot, however, in its Eq. (13) form be used for low temperature spectra.

2.7. Selection Rules from the Judd-Ofelt Theory

Several selection rules may be obtained from the Judd-Ofelt Theory. The $6\text{-}j$ symbol in Eq. (10) gives $\lambda \leqslant 6$ (and so because λ is even, $\lambda = 2,4,6$); $t \leqslant 7$ (and so because t is odd, $t = 1,3,5,7$) and $\Delta l = \pm 1$ (which means that only $l = d$ and g configurations may be mixed into $4f^N$). The $3\text{-}j$ symbol in Eq. (9) give the selection rules on t with respect to λ: $\lambda = 2$, $t = 1,3$; $\lambda = 4$, $t = 3,5$; $\lambda = 6$, $t = 5,7$. If the reduced matrix elements $\Gamma^{(\lambda)}$ are further reduced by means of Eq. (2), selection rules on S, L and J are obtained. The Kronecker delta gives $\Delta S = 0$ and the $6\text{-}j$ symbol $|\Delta J| \leqslant \lambda$, $|\Delta L| \leqslant \lambda$, and if J or $J' = 0$, $|\Delta J|$ must be even, and $J = 0 \leftrightarrow J' = 0$ is forbidden. Finally the $3\text{-}j$ symbol involving M and M' in Eq. (11) gives selection rules on M and M' (which in turn give selection rules on the crystal field quantum numbers μ): $|\Delta M| = p + q$; so for σ polarised spectra $(q = 0)$ $|\Delta M| = 3p$ and for π polarised spectra $(q = \pm 1)$ $|\Delta M| = p \pm 1$. The value of p is of course determined by the particular point group considered: for example in D_{3h} symmetry the odd crystal field parameters are (11) A_{33}, A_{53} and A_{73} thus $|\Delta M| = 3$ (σ) and $|\Delta M| = 2,4$ (π). Summarising then:

$$\Delta l = \pm 1 \qquad \qquad \Delta S = 0 \qquad \qquad |\Delta L| \leqslant 6$$

$$|\Delta J| \leqslant 6 \quad \text{unless } J \text{ or } J' = 0 \quad \text{when } |\Delta J| = 2,4,6.$$

$$|\Delta M| = p + q$$

The selection rules on S and L are valid in the limit of Russell-Saunders coupling, but since transitions are between linear combinations of Russell-Saunders states they are not rigidly adhered to. Thus the $^5D_2 \leftarrow {}^7F_0$ transition of Eu(III) is allowed because the state labelled $^5D_2$ is a linear combination of $^5D_2$, $^7F_2$ etc. The selection rules on J however are much more rigid and can be broken only by 'J-mixing' which is a rather weak effect. For example the $^6F_{1/2} \leftarrow {}^6H_{15/2}$ $(\Delta J = 7)$

transition is missing from most Dy^{3+} spectra *(11)* and the $^4F_1 \leftarrow {}^4I_8$ $(\Delta J = 7)$ transition could not be identified in the spectrum of Ho^{3+} in $YAlO_3$ *(18)*. The selection rules which apply when J or $J' = 0$ are exemplified by the absence (or extreme weakness) of $^5D_{3\ 0} \leftarrow {}^7F_0$ transitions of Eu(III). In fact the $0 \leftrightarrow 0$ rule is broken in many compounds of Eu(III) and this will be discussed in Section 7.

3. Other Multipole Transitions

As well as electric dipole radiation, $f \leftrightarrow f$ transitions can absorb magnetic dipole and higher even electric multipole radiation (quadrupole, hexadecapole and 64-pole). The oscillator strength of a magnetic dipole transition is given by

$$P_{\text{M.D.}} = \chi \left[\frac{2 \pi^2}{3 h m c} \right] \langle f^N \psi J M \, | \mathbf{L} + g_i \mathbf{S} | f^N \psi' J' M' \rangle^2 \tag{16}$$

If the matrix element is reduced and we average over all values of M (assuming equal population) and of the components of $\mathbf{L} + g_i \mathbf{S}$ we have for isotropic spectra:

$$P_{\text{M.D.}} = \chi \left[\frac{2 \pi^2}{3 h m c} \right] \langle f^N \psi J \| \mathbf{L} + g_i \mathbf{S} \| f^N \psi' J' \rangle^2 (2J + 1)^{-1} \tag{17}$$

The reduced matrix elements may be written as closed formulae and explicitly evaluated. The selection rules are:

$$\Delta l = 0 \qquad \Delta S = 0 \quad \Delta L = 0 \qquad \Delta J = 0, \pm 1 \quad (\text{not } 0 \leftrightarrow 0)$$
$$\Delta M = 0 \qquad (\sigma \text{ polarisation})$$
$$\Delta M = \pm 1 \ (\pi \text{ polarisation})$$

In the limit of Russel-Saunders coupling, transitions can occur only between J-levels of the ground term and so most strong magnetic dipole transitions are outside the available spectral range. Due to the relaxation of the S and L selection rules by intermediate coupling some magnetic dipole transitions do occur in the visible region. They are in general at least an order of magnitude less intense than electric dipole transitions and are only seen when the latter are very weak. The $^5D_1 \leftarrow {}^7F_0$ and $^5D_0 \rightarrow {}^7F_1$ transitions of Eu(III) have been established *(19)* as predominantly magnetic dipole by polarisation measurements as have the $^6P_{7/2} \leftarrow {}^8S_{7/2}$ and $^6P_{5/2} \leftarrow {}^8S_{7/2}$ transitions of Gd(III) *(20)*. *Carnall et al.* have published a full list *(17)* of calculated magnetic dipole oscillator strengths.

Electric quadrupole transitions are calculated to be weaker still. The oscillator strength is given by *(11)*:

$$P_{\text{E.Q.}} = \chi \left[\frac{8 \pi^2 m c}{3 h} \right] \sigma \, \Omega_2(\text{QUAD.}) \, (\Gamma^{(2)})^2 \, (2J + 1)^{-1} \tag{18a}$$

where

$$\Omega_2 (\text{QUAD.}) = \sigma^2 \left(\frac{2 \pi^{(2)}}{15} \right) \langle f \| C^{(2)} \| f \rangle^2 \langle 4f | r^2 | 4f \rangle^2 \tag{18b}$$

93

No electric quadrupole $f \leftrightarrow f$ transitions have ever been experimentally identified (19). They would obey the selection rules $\Delta l = 0$, $|\Delta J|$ and $|\Delta L| \leqslant 2$, $\Delta S = 0$. Any transition characterised by a $\Gamma^{(2)}$ reduced matrix element is of course potentially electric quadrupole. Higher multipole transitions are calculated to have negligible intensity.

4. The Success of the Judd-Ofelt Theory

4.1. Obtaining the Ω_λ Parameters from Experimental Oscillator Strengths

It has become customary to treat the Ω_λ as phenomenological parameters to be obtained from the experimental oscillator strengths and the calculated reduced matrix elements via Eq. (12) or (13). It has beeen shown (21) that the values of the reduced matrix elements are not very sensitive to the particular ligands surrounding the lanthanide ion being considered, and a comparison of the various sets which have been published shows that the differences between them are negligible compared to the other errors involved. (The dependence of the reduced matrix elements upon the environment of the ion comes of course from the intermediate coupling coefficients which in turn are functions of the electrostatic energy and spin-orbit parameters). Most studies of the solution spectra of lanthanide complexes and some treatments of ions in crystals have used the excellent sets of reduced matrix elements calculated by *Carnall et al.* (22) for the aquo ions.

The procedure then is to solve the equation

$$P = \chi \left[\frac{8\pi^2 m c}{3 h} \right] \sigma \sum_{\lambda=2,4,6} \Omega_\lambda (\Gamma^{(\lambda)})^2 (2J+1)^{-1} \tag{13a}$$

for the Ω_λ parameters by a least squares procedure. Most authors have then calculated the oscillator strength from these values and expressed the difference between the observed and calculated values of P in terms of the root mean square deviation (RMS) defined as:

$$RMS = \left(\frac{\text{sum of squares of deviations}}{\text{number of observations-number of parameters}} \right)^{\frac{1}{2}}$$

Some authors have used alternative statistical methods (23) to solve Eq. (13a) but there is no essential difference. This treatment has been applied both to solution spectra and to those systems of ions in crystals where the individual Stark components cannot be resolved. Where the Stark components can be resolved Eq. (11) can be used. In this case the symmetry of the ion site determines which of the A_{tp} are non-zero and Eq. (11) may be solved in the same way as Eq. (13a) treating $A_{tp} \Xi(t, \lambda)$ as phenomenological parameters.

Ω_λ (or $\mathcal{T}_\lambda$) parameters have now been obtained for quite a number of systems. The most complete work is that by *Carnall et al.* (16, 17) on the aquo ions where $\mathcal{T}_\lambda$ parameters for all the aquo ions (including radioactive Pm^{3+}) are presented.

In the same paper *(16)* certain lanthanide nitrates in ethyl acetate (Pr^{3+}, Nd^{3+}, Er^{3+}, Tm^{3+} and Yb^{3+}) and nitrates in $LiNO_3$-KNO_3 eutectic at 160 °C (Pr^{3+}, Nd^{3+}, Er^{3+}, Tm^{3+} and Yb^{3+}) are treated. The present author has considered *(24, 25)* the decatungstolanthanate(III) series for Pr^{3+}, Nd^{3+}, Sm^{3+}, Eu^{3+}, Dy^{3+}, Er^{3+} and Tm^{3+}, certain other heteropolytungstate complexes of Pr^{3+} and Ho^{3+} *(24)* and has investigated the sensitivity of the $\mathscr{T}_\lambda$ parameters to the particular transitions used in the data set *(26)*. Tris β-diketonate complexes of Pr^{3+}, Nd^{3+} and Er^{3+} have been investigated *(27, 28, 29)* as have labile complexes of Ln^{3+} with α-picoline *(23)* (Pr^{3+}, Nd^{3+}, Sm^{3+}, Dy^{3+}, Ho^{3+}, Er^{3+} and Tm^{3+}), acetate *(30)* (Pr^{3+} and Nd^{3+}), aromatic acids *(31)* (Eu^{3+}) and a number of nitrogen-co-ordinating ligands *(32)* (Nd^{3+}). The gaseous tribromides and tri-iodides of Pr^{3+}, Nd^{3+}, Er^{3+} and Tm^{3+} have been discussed *(33, 34)* by *Gruen, De Kock* and *McBeth*. Ω_λ parameters have been obtained for ions doped in the following hosts: Y_2O_3 *(35, 36)* (Pr^{3+}, Nd^{3+}, Eu^{3+}, Er^{3+}, Tm^{3+}), LaF_3 *(18, 35, 37)* (Pr^{3+}, Nd^{3+} and Er^{3+}), yttrium aluminium garnet *(38)* (Nd^{3+}), $YAlO_3$ *(39, 40)* (Pr^{3+}, Nd^{3+}, Eu^{3+}, Tb^{3+}, Ho^{3+}, Er^{3+} and Tm^{3+}), CaF_2 *(20)* (Gd^{3+}) and Eq. (11) has been used to obtain sets of $A_{tp}\ \Xi\ (t, \lambda)$ for the following systems: europium *(41)* and thulium *(42)* ethylsulphates and Ho^{3+} doped in YPO_4 *(43)*.

The largest number of (and probably the most accurate) Ω_λ parameters have been obtained for compounds of Nd(III) and in Table 1 are collected the parameters for a number of systems along with the site symmetry of the Nd^{3+} ion (where known) and the RMS value.

The values of the Ω_λ parameters are rather sensitive to the accuracy of the oscillator strengths used to compute them. The measurement of the oscillator strengths of solution spectra presents no difficulty. Much of the work on ions in crystals, however, has been done on doped samples where the concentration is known to only $\pm 10\%$ and, more important, the weaker transitions in a spectrum have often been measured using more highly doped samples than those used for the stronger transitions; thus introducing a rather larger relative error in the oscillator strengths of different transitions within a system. Further complications arise when lanthanide ions can enter more than one site in the host (*e.g.* in Y_2O_3 there are sites of both C_2 and S_6 symmetry and Ln^{3+} ions are known to enter both) since analyses give only the total concentration of lanthanide in the crystal. Another source of error in crystal systems which is not encountered in solution work is the fact that the refractive index (and so χ) is frequency dependent and has in certain cases been measured with low accuracy, or not at all, in parts of the spectral range. In view of the foregoing it is not surprising to find that the RMS values obtained for doped systems are generally larger than those for well-defined complexes in solution. It should be noted, however, that many lanthanide complexes are labile in solution. Although this should not affect the accuracy of the Ω_λ parameters it will affect the interpretation put upon them since they will be made up of the Ω_λ's of the various species present weighted by their equilibrium concentrations. Even complexes with β-diketones can be substantially dissociated in co-ordinating solvents if the concentration is low *(45)*.

The values of Ω_λ obtained also depend to some extent upon the particular transitions included in the data set *(26)*. From Eq. (13a) it can be seen that in the fitting procedure each Ω_λ is weighted with the value of the reduced matrix

Table 1. Judd-Ofelt parameters for compounds of Nd(III)

Ω_λ ($\times 10^{20}$) cm²	2	4	6	RMS. ($\times 10^7$)	Point group	Ref.
Nd³⁺ aquo	0.93 ± 0.3	5.0 ± 0.3	7.9 ± 0.4	5.8	?	(17)
Nd(NO₃)₃ in ethyl-acetate	9.2 ± 0.4	5.4 ± 0.3	7.7 ± 0.45	6.6	?	(16)
Nd(NO₃) in LiNO₃/KNO₃	11.2 ± 0.3	2.6 ± 0.3	4.1 ± 0.4	6.1	C_3?	(16)
K₇[NdW₁₀O₃₅]	0.8 ± 0.45	5.6 ± 0.6	7.0 ± 0.3	3.6	D_{4d}	(25)
Nd(α-opicolinate)	3.2 ± 0.85	10.4 ± 1.2	10.3 ± 0.6	15.6	?	(23)
Nd(acetylacetonate)₃ in DMF	24.5	0.71	9.1	3.0	C_3–D_3	(32)
Nd(acetylacetonate)₃ in MeOH/EtOH	15.7	0.73	7.4	2.5	C_3–D_3	(32)
Nd(dibenzoylacetyl-acetonate)₃ in MeOH/EtOH	34.1	2.5	9.1	3.4	C_3–D_3	(32)
NdBr₃	180	9	9	—	D_{3h}	(33)
NdI₃	275	9	9	—	D_{3h}	(33)
Nd³⁺ in yttrium aluminium garnet	0.2	2.7	5.0	—	D_2 or D_{2d}	(38)
Nd³⁺ in Y₂O₃	8.6 ± 0.4	5.3 ± 0.8	2.9 ± 0.6	7.8	C_2	(35)
Nd³⁺ in LaF₃	0.35 ± 0.1	2.6 ± 0.4	2.5 ± 0.3	4.0	C_2	(35)
Nd³⁺ in YA lO₃	1.3	4.7	5.7	6.0	C_{1h} or D_{3h}	(39—40)
NdCl₆³⁻	~0.7	~.5	~.8	—	~O_h	(52)
NdBr₆³⁻	~1.0	~.5	~.8	—	~O_h	(52)

element associated with it. An inspection of the tables of reduced matrix elements in Ref. (22) shows that, in general, a large number of transitions in the accessible spectral region will have relatively large values of $\Gamma^{(6)}$, less will have large values of $\Gamma^{(4)}$ and relatively few (usually two, one or even none) will have large values of $\Gamma^{(2)}$. Thus the Ω_6 parameters will usually be the best defined and Ω_2 the least. In those cases where there are no transitions with large values of $\Gamma^{(2)}$ the Ω_2 parameter will become a 'fitting parameter' varying randomly and sometimes even taking a negative value (which is not possible within the framework of the Judd-Ofelt theory). Ω_4 may, less often, also act as a fitting parameter. Examples of cases where the value of Ω_2 is probably meaningless are the Tb³⁺ aquo ion (.004 × 10⁻²⁰ cm²), [SmW₁₀O₃₅]⁷⁻ (121 × 10⁻²⁰ cm²), [DyW₁₀O₃₅]⁷⁻ (−23 × 10⁻²⁰ cm²) and Sm³⁺ α-picolinate (83 × 10⁻²⁰ cm²); Ω_4 is probably acting as a fitting parameter to some extent in Tm³⁺ α-picolinate.

One further point regarding the determined values of the Ω_λ parameters may be mentioned. It has been found (26) that when more than one transition with a large value of $\Gamma^{(2)}$ is included in the data set a much better fit is obtained if each transition is independently assigned an Ω_2 parameter. Since this effect is not found for Ω_4 and Ω_6 it is thought that the improvement in the fit is not due merely to the presence of extra independent variables. Transitions characterised by large values of the $\Gamma^{(2)}$ matrix element are in general rather sensitive to the environment of the lanthanide ion and have been called (46) 'hypersensitive' by *Jørgensen* and *Judd*. They are discussed in detail in Section 5. For the rest of this section we will be mainly concerned with the Ω_4 and Ω_6 parameters. Since Ω_6 is usually better defined than Ω_4 the former will be most often mentioned but all conclusions about Ω_6 will also apply to Ω_4.

The phenomenological treatment has been rather successful in as much as it can account for the intensities of, say, twenty transitions whose relative intensity may vary over as much as two orders of magnitude with an agreement which is usually better than 10% of the average oscillator strength. This has been taken as a vindication of the Judd-Ofelt theory when really it shows only that the formalism of representing the intensity of an $f \leftrightarrow f$ transition by the sum of (phenomenological parameters multiplied by the squares of the non-zero, even $\lambda, \Gamma^{(\lambda)}$ matrix elements) is useful. To put it another way, the success of Eq. (13a) implies nothing about the definition of the Ω_λ parameters and so nothing about the physical model. It will be shown later, for example, (Section 5.7) that the equation

$$ P = \chi \left[\frac{8 \pi^2 m c}{3 h} \right] \sigma \, \Omega_2 (\Gamma^{(2)})^2 \, (2J+1)^{-1} $$

can be derived from an entirely different physical model from that used in the Judd-Ofelt theory. Ω_2 in this case has a quite separate meaning from the Ω_2 occurring in Eq. (13a). In order then to test the Judd-Ofelt theory it is necessary to consider the experimental values (and trends among such values) of the Ω_λ parameters and see how they compare with the predictions of Eq. (13b).

Firstly it can be noted that Ω_λ will contain both static and vibronic contributions. It is rather difficult to say much about the expected behaviour of Ω_λ (vibronic) so it will be helpful to assess the relative contributions from the two mechanisms. The Ω_λ (static) can then be considered in a number of ways: a comparison of Ω_λ for the same element in different environments or the variation of Ω_λ across the series of Ln^{3+} ions in the same host may usefully be studied. Finally the parameters may be calculated and compared with experiment.

It should be mentioned at this point that the Pr^{3+} ion has been much less successfully treated by the phenomenological method than have the other lanthanide ions. It is possible that an additional intensity mechanism is operating and the intensities of Pr^{3+} spectra will be treated separately in Section 6.

4.2. An Assessment of the Relative Static and Vibronic Contributions to Ω_λ

The Ω_λ parameters, as phenomenologically obtained, are a combination of Ω_λ due to a static perturbation and Ω_λ due to a vibronic perturbation (as well as possible contributions from a quadrupole or augmented quadrupole mechanism

97

as discussed in Section 4.5). There are three ways in which intensity arising from these two perturbation mechanisms may be distinguished.

a) It may be possible, particularly in crystal spectra at low temperatures, to distinguish the vibronic structure from the pure electronic lines.

b) We may investigate complexes possessing a centre of symmetry where all the intensity must be due to a vibronic mechanism.

c) The temperature dependence of the intensity and so of Ω_λ may distinguish between the two mechanisms since only Ω_λ (vibronic) should be temperature sensitive [through $\varrho(\eta)$ of Eq. (15)].

Vibronic structure has been observed in the crystal spectra of a large number of lanthanide compounds. The proportion of vibronic to electronic intensity varys with the ligands or crystal host and also with the particular lanthanide ion considered. It seems (35) that in general the vibronic contribution is larger at the beginning and end of the series (i.e. Pr^{3+}, Nd^{3+}, Tm^{3+}, Yb^{3+}) than in the middle. Thus in the case of Pr^{3+} in $LaCl_3$ studied by *Satten et al.* (47) more than 50% of the intensity of the $^3P_0 \, (\mu = 0) \leftarrow {}^3H_4(\mu \pm 2)$ transition was found to be due to vibronic lines. On the other hand a recent study (48) of the absorption and emission spectra of Eu^{3+} in a number of tungstate hosts, in which the vibronic structure is very well resolved, shows that the relative proportion of vibronic to static intensity is between 10 and 20%. It has been known for a considerable time that vibronic structure can be observed in solution spectra under favourable conditions. *Freed et al.* (49) found vibronic structure in the spectrum of $EuCl_3$ in ethanol but no relative intensities were quoted. The fluorescence spectrum of Gd^{3+} perchlorate in water or D_2O (50) shows, to lower energy of the $^6P_{7/2} \rightarrow {}^8S_{7/2}$ transition, side bands which have been interpreted as vibronic transitions: the frequency difference being that of the O—H or O—D stretch of the co-ordinated H_2O or D_2O molecules. The intensity of these bands is about 1/60 of the electronic intensity. The spectra of many lanthanide compounds, especially discrete molecular complexes, show no vibronic structure, even under high resolution. Notable cases are a large number of Eu^{3+} trisβ-diketonates studied at 70 K by *Blanc* and *Ross* (51).

The number of strictly centrosymmetric compounds of the lanthanides is rather limited. The hexahalides Ln X_6^{3-} (X = Cl, Br, I) have absorption spectra which are in general an order of magnitude less intense than the corresponding aquo ions (52, 53). The hypersensitive transitions are exceptions to this statement having intensities but little less than the aquo ion transitions. These spectra have been interpreted (52) as showing that the hexahalides are octahedral (or nearly octahedral), this conclusion being supported by the 4 K absorption spectrum of a single crystal of $[(C_6H_5)_3PH]_3 \, NdCl_6$ (54). The published spectrum (52) of the $^3P_{2,1,0}$, $^1I_6 \leftarrow {}^3H_4$ multiplet of $PrCl_6^{3-}$ shows that the oscillator strengths due to the vibronic transitions are considerably less than 10% of the intensity of these transitions in the Pr^{3+} aquo ion and leads to the conclusion that even for those lanthanides where considerable vibronic intensity is found in a crystal lattice environment little vibronic intensity is found in molecular complexes.

A similar intensity situation is reported for the hexakisisothiocyanatolanthanate (III) series (55) where the Er(III) complex has been shown by X-ray crystallography to be octahedrally co-ordinated. [It should be noted however that what appear superficially to be the same complexes have been prepared and reported by another group of workers (56). In this case the intensities are much larger and are comparable to those of the aquo ions]. The spectra of the hexanitrolanthanates(III) (57) and those of Er(III) and Nd(III)hexakisantipyrenetriiodides, (58) in all of which the lanthanide is octahedrally co-ordinated, appear to be considerably weaker than the corresponding aquo ion spectra. More conclusive perhaps is a study by *Blasse, Bril,* and *Nieuwport* (59). They have prepared a number of phosphors containing Eu(III) in which the europium ion has a precisely known site symmetry. In two cases (Ba_2GdNbO_6:Eu and Gd_2TiO_7: Eu) the Eu^{3+} is at a centre of inversion and the fluorescence spectrum shows that while the $^5D_0 \rightarrow$ $^7F_1$ magnetic dipole transition has an intensity comparable to that exhibited at other symmetries, the $^5D_0 \rightarrow ^7F_2$ transition is very weak (for example it has less than 5% of the intensity of the same transition in $YAl_3B_4O_{12}$: Eu where the symmetry of the Eu^{3+} is D_{3h}).

Temperature variation studies of the intensities of lanthanide spectra are complicated by the fact that, except for Eu(III), the ground J-states are split by the crystal field into a number of levels (called Stark components). The splitting can be as large as 600 cm^{-1} so the Stark components are unequally populated and will change their population with temperature. In order to allow for this it is necessary to know the energy and crystal field quantum number of each component; such information is available for few systems and for only two whose intensities have been studied. These are: $[(C_6H_5)_3PH]_3NdCl_6$ where it was found (54) that the intensities of the individual Stark transitions of all the accessible absorption bands are temperature independent and Tm^{3+} ethylsulphate where Ω_λ parameters were obtained (42) at 300 K and 23 K. Ω_4 and Ω_6 were found to be very much larger at the higher temperature implying a considerable vibronic contribution in this case. Europium is much easier to deal with. Here the ground level ($^7F_0$) is unsplit by the crystal field and although there are other J-levels at thermally accessible energies their positions and identities are well known and can be allowed for. The spectrum of a solid glass of $K_{13}[Eu(SiW_{11}O_{39})_2]$ has been measured (60) at 300 K and 80 K and it has been shown that the oscillator strengths of transitions to the $^5D_1$, $^5D_2$, $^5L_5$ and $^5G_{4,5,6}$ levels are all temperature invariant. In the same study (60b) it was shown that the intensity of all the visible bands of the isomorphous Ho^{3+} complex increased by about 10% on cooling. Analogy with transition-metal ion complexes suggests that the intensity of a vibronically allowed electric dipole transition should decrease by 30—40% on cooling from 300 to 80 K and so it seems again that a vibronic mechanism is unimportant for molecular complexes.

In conclusion, the Ω_λ parameters have been shown by several studies to consist of both vibronic and static parts. The former can be very large for certain ions in crystals but is considerably less important for molecular complexes of the lanthanides. It has been suggested (60b) that lattice vibrations are more effective in mixing excited configurations into the $4f^N$ configuration than are the higher frequency metal-ligand vibrations available in molecular complexes.

4.3. The Variation of Ω_λ with the Environment of the Lanthanide Ion

For a given lanthanide ion the variation of Ω_λ with environment is determined by the crystal field parameters, A_{tp}. These in turn depend upon the positions (angular dispositions and distance) and the charges of the surrounding ligands. Unfortunately inspection of Table 1 shows that the precise site symmetry (let alone structure) is known for only a few systems and most of these are cases of ions doped in crystals where due to uncertainties of concentration and multiple site occupancy the values of Ω_λ are most suspect. It can be seen at once however that Ω_2 varies enormously with environment (this is discussed in Section 5) while Ω_4 and Ω_6 are generally within a factor of 3 of each other. The one exception to this are the hexahalides where Ω_4 and Ω_6 are about an order of magnitude less than the aquo ion values and much of this is due to the vibronic contribution. This is quite consistant with the behaviour expected for an octahedral environment since no A_{tp} with t odd are permitted in the O_h crystal field expansion. The relative insensitivity of Ω_4 and Ω_6 to environment is, however, rather worrying since it is well known (67) that A_{tp}, at least when calculated from a point charge model, are very sensitive to the positions of the ligands and we would not expect 3-co-ordinate $NdBr_3$, 6-coordinate $Nd(acetylacetonate)_3$ and the 9-co-ordinate Nd^{3+} aquo ion to have very similar A_{tp}. The electrostatic model also predicts that Ω_λ should be proportional to $R^{-(2t+2)}$ where R is the distance from the lanthanide ion to the ligand and $t=3$ and 5 for $\lambda=4$ and 5 and 7 for $\lambda=6$. Yet while the $Ln–OH_2$, $Ln–Br$ and $Ln–I$ bond lengths are quite different the respective $\Omega_{4,6}$ are very similar. This leads to the suggestion that the perturbation may not be a crystal field (at least in the simple electrostatic charge sense). This is further discussed in Section 4.5. For the rest, little more can be said except to note that the insensitivity of Ω_4 and Ω_6 to environment makes the intensities of the non-hypersensitive transitions of little use in solving structural problems.

4.4. The variation of Ω_λ with Number of f Electrons

If we restrict ourselves to an isostructural series of complexes or to a series of ions doped into the same host we have effectively removed one of the variables of Eq. (13b): the crystal field parameters. The expression for Ω_λ then becomes:

$$\Omega_\lambda = \sum_{p,t} C_{pt} \sum_t \Xi(t,\lambda)$$

where C_{pt} are now a set of constants and are functions of the structure of the host. Assuming that the only perturbing configurations are $4f^{N-1}5d$ and $4f^{N-1}ng$ the values of t which are non zero may be obtained by considering the $\langle l \| C^{(t)} \| f \rangle$ matrix element of Eq. (10). For $4f^{N-1}5d$ $(l=2)$ $t \leq 5$; for $4f^{N-1}ng$ $(l=4)$ $t \leq 7$. The 6-j symbol then gives the values of t associated with each value of λ. Noting that t must be odd we have finally that the only non-zero $\Xi(t,\lambda)$ are $\Xi(1,2)$, (3,2) (3,4), (5,4), (5,6) and (7,6). The value of $\Xi(t,\lambda)$ can be calculated via Eq. (10) from published values of the radial integrals and configuration energies (although interpolation is required for the ions in the middle of the lanthanide series).

The six $\Xi\,(t, \lambda)$ parameters listed above have been calculated by *Krupke* (*35*) for Pr^{3+}, Nd^{3+}, Eu^{3+} Tb^{3+}, Er^{3+}, and Tm^{3+}. In each case the value of $\Xi\,(t, \lambda)$ decreases monotonically from Pr^{3+} to Tm^{3+}. Since the sum of a series of monatonic curves must itself be a monotonic curve it follows that the Judd-Ofelt theory predicts that the variation of Ω_λ across an isostructural series of lanthanide ions should be monotonic: this conclusion holding only if the intensity is produced by a static perturbation. The experimental variation of Ω_λ across a number of isostructural series will now be considered.

Ω_2. As might be expected, the data for this parameter is sparse and conflicting. Values for the isostructural series $[LnW_{10}O_{35}]^{7-}$ appear to vary rather randomly (*25*) as do these for Ln^{3+} doped in $YAlO_3$ (*39*) and LaF_3 (*18, 35, 37*). The variation in the aquo ion series (*17*) (which it is believed (*25*) are isostructural in dilute solution) is also irregular. In contrast Ω_2 for Ln^{3+} doped in Y_2O_3 (*35, 36*) appears to decrease monotonically from Nd^{3+} to Tm^{3+} although the spectrum of the Ho^{3+} ion was not obtained and in both $[LnW_{10}O_{35}]^{7-}$ and the aquo ions Ω_2 for Ho^{3+} is about half that for Er^{3+} making a regular variation difficult.

Ω_4 and Ω_6. Here the data is much more reliable and clear trends can be seen. Figs. 1, 2 and 3 show the variation of Ω_6 with number of f-electrons for $[LnW_{10}O_{35}]^{7-}$, the Ln^{3+} aquo ions and for Ln^{3+} doped in Y_2O_3 respectively.

In the first two series the variation is virtually linear [reasons why Pm^{3+} and Sm^{3+} might be irregular are discussed in Ref. (*25*)] thus confirming the prediction of the Judd-Ofelt theory. Similar behaviour is exhibited by the Ω_6 parameters of Ln^{3+} doped in $YAlO_3$ and in LaF_3; in these cases the points lying on a shallow curve.

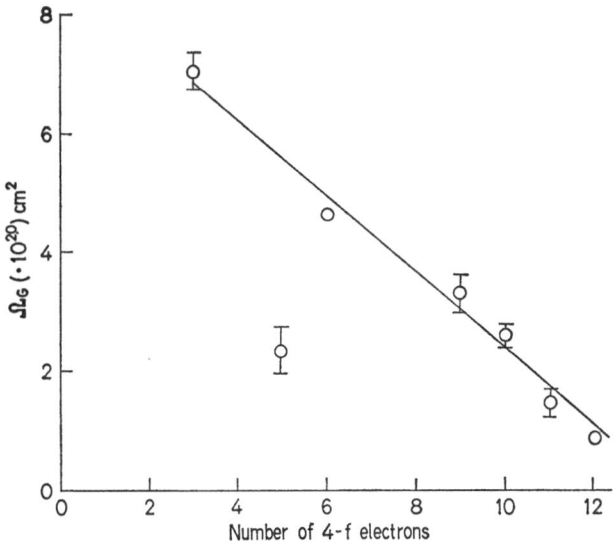

Fig. 1. Variation of the Ω_6 parameters of $[LnW_{10}O_{35}]^{7-}$ with number of $4f$ electrons [data from Ref. (*25*)]

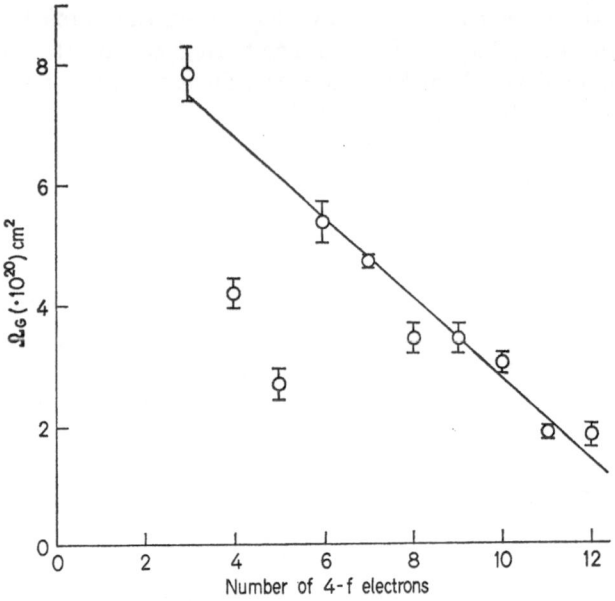

Fig. 2. Variation of the Ω_6 parameters of Ln^{3+} aquo ions with number of $4f$ electrons [data from Ref. (*17*)]

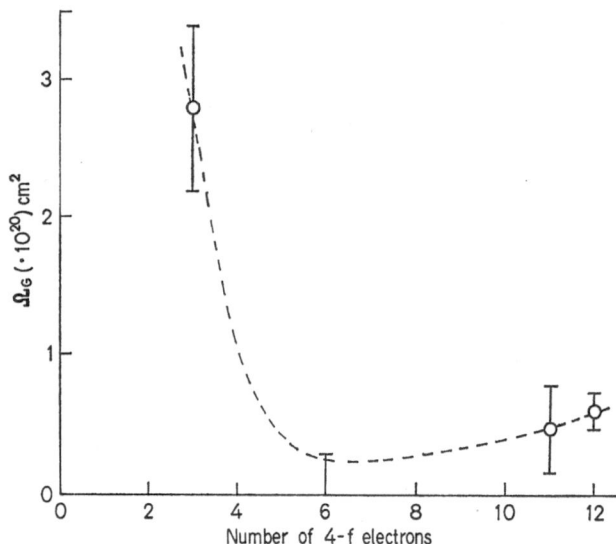

Fig. 3. Variation of the Ω_6 parameters of Ln^{3+} doped in Y_2O_3 with number of $4f$ electrons [data from Ref. (*35*)]

For Ln^{3+} doped in Y_2O_3 the picture is somewhat different. The value of Ω_6 goes through a minimum at Eu^{3+}. The explanation of this appears to be that both at the beginning and end of the series there is a substantial contribution to Ω_6 from the vibronic mechanism which is supported by the observation (35) of vibronic structure in the spectra of Nd^{3+} and Tm^{3+} but none in the Eu^{3+} spectrum. This agrees with the conclusion reached in the previous section about the relative importance of the two perturbation mechanisms in molecular complexes and in ionic crystals. In all cases mentioned above the Ω_4 parameters exhibit the same behaviour as do Ω_6 but since they are in general less accurately known the scatter of points is somewhat greater.

4.5. Calculations of the Ω_λ Parameters

Two attempts have been made to calculate the Ω_λ parameters; one by *Judd* (6) for the aquo ions and one by *Krupke* (35) for the Y_2O_3 host. In both cases the crystal field parameters were calculated from a point charge model and values of the radial integrals were obtained by interpolation between those for Pr^{3+} and Tm^{3+} which have been calculated (62) by *Rajnak*. In considering the aquo ions *Judd* used as his model the known structure of $GdCl_3 \cdot 6H_2O$ with the two Cl^- ions removed from the co-ordination polyhedron of Gd^{3+} leaving it surrounded by six water molecules. Including $4f^{N-1}5d$ and all $4f^{N-1}ng$ configurations (but not $4f^{N-1}nd$ with $n \geqslant 6$ which he argued would make a negligible contribution) he calculated Ω_λ values which were 2—3 times too small for Nd^{3+} and 5—10 times too small for Er^{3+}. A substantial contribution to the parameters came from the g electron configurations despite their high energy.

Krupke's calculation, using the known crystal structure of Y_2O_3 might be expected to give better agreement. Considering the results for Eu^{3+} which is least complicated by a contribution from Ω_λ (vibronic) the calculation overestimated Ω_2 by a factor of 2 and Ω_4 by factor of 10 while underestimating Ω_6 by a factor of 10. Again the g orbital contribution was very significant. *Krupke* rationalised the discrepancy by postulating the need to adjust the values of the radial integrals. By reducing the value of $\langle 4f|r|5d \rangle$ and increasing that of $\langle 4f|r^8|4f \rangle$ (which gives an increase in the value of $\langle 4f|r^7|ng \rangle$ via the relationship $\sum_n \langle 4f|r|ng \rangle \langle ng|r^t|4f \rangle =$ $\langle 4f|r^{t+1}|4f \rangle$) a much better agreement with experiment was obtained. In doing so the g orbital contribution to the parameters was increased still further to 80% (Ω_2), 50% (Ω_4) and 100% (Ω_6).

While appreciating that the values of radial integrals may have to be modified from their free ions values and that Krupke's arguments may indeed be valid, the present author feels that a much more likely source of the discrepancy between the calculated and experimentally determined values of Ω_λ can be found in the practice of calculating the crystal field parameters from an electrostatic point charge model. A recent review (63) of the theory of the lanthanide crystal field by *Newman* states clearly that crystal field parameters calculated from a point charge model differ very substantially from the experimental values. By including other factors such as covalency and configuration interaction much better agreement can be obtained. It is instructive to consider the contributions to the crystal field parameters of $PrCl_3$ from the various mechanisms. These are given in Table 12 of Ref. (63) along with the experimental parameters. If only the point charge part of the calculation is compared with experiment the $t = 2$ parameter is found to be greatly overestimated, the $t = 4$ one is approximately correct and the $t=6$ parameter is very considerably underestimated. If this trend also applies to A_{tp} with t odd we may be able to understand why Krupke's calculation overestimated Ω_2 and Ω_4 and underestimated Ω_6. In view of these arguments it is probably safe to say that a comparison of Ω_λ with calculations based on a electrostatic point charge model is no test of the theory. A better approach may be to obtain odd parity crystal field parameters from intensity data and then try and calculate these with the various theories available. One attempt (43) at obtaining crystal field parameters from intensity measurements has been made.

4.6. Conclusion

The Judd-Ofelt theory has obviously been a great help in understanding lanthanide intensities. Instead of having to deal with a large number of individual transitions it is now possible to consider the behaviour of only three parameters for each lanthanide. Although the theory has not really been subjected to sufficient test it appears to be a good working model, at least for the non-hypersensitive transitions, provided the perturbation is not considered too literally to be due to the electrostatic crystal field of the ligands.

5. Hypersensitive Transitions

5.1. Introduction

Most of the $f \leftrightarrow f$ transitions of the trivalent lanthanides have intensities which are little affected by the environment of the ion. A few, however, are very sensitive to the environment and are usually more intense when the ion is complexed than they are in the corresponding aquo ions (which for historical reasons have been taken as a standard with which other compounds are compared). Such transitions have been called hypersensitive transitions (46) by *Jørgensen* and *Judd*. Spectra of aqueous solutions of Nd^{3+}, Eu^{3+}, Ho^{3+} and Er^{3+} nitrates of varying concentration obtained by *Selwood* (64) in 1930 show clear hypersensitivity associated with certain absorption bands although the phenomenon was not commented upon at the time. *Freed* noticed (65) that the intensity of the $^5D_2 \leftarrow {}^7F_0$ transition of Eu^{3+} was sensitive to the environment of the Eu^{3+} ion, but not until *Moeller* and his co-workers (66) began studying the complexes of Nd^{3+}, Ho^{3+} and Er^{3+} with EDTA. and β-diketones was hypersensitivity realised to be a general phenomenon. Figure 4 shows the spectrum of a typical hypersensitive complex of Nd^{3+} along with that of the Nd^{3+} aquo ion.

Shortly after *Judd*'s paper (6) on the intensities of $f \leftrightarrow f$ spectra, he and *Jørgensen* considered (46) the problem of hypersensitivity. They noted that all known hypersensitive transitions obeyed the selection rules $|\Delta J| \leqslant 2$, $|\Delta L| \leqslant 2$ and $\Delta S = 0$ and that these were just the selection rules on $\Gamma^{(2)}$. This suggested that hypersensitivity could then be rationalised as the peculiar sensitivity of Ω_2 to the environment. Since then $\Gamma^{(\lambda)}$ for all the lanthanide elements have been published (32). It can now be said that all transitions in accessible spectral regions whose intensity is dominated by the value of $\Gamma^{(2)}$ are hypersensitive (apart from Gd^{3+} where no data except that for the aquo ions seems to be available). Table 2 lists

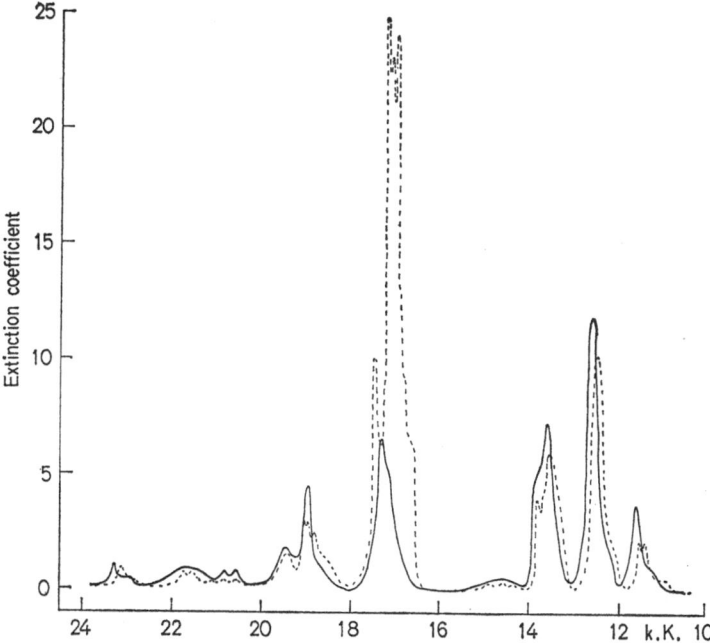

Fig. 4. The solution spectrum (in $CHCl_3$) of Nd(hydroxymethylenecamphorato)$_3$ (broken line) and the aqueous solution spectrum of Nd^{3+} perchlorate (full line) [from Ref. (*112*)]

Table 2. Hypersensitive transitions

Lanthanide	Transition[a]	Approximate wavenumber ($\times 10^{-3}$) cm^{-1}	Ref.[b]
Pr(III)	$^3F_2 \leftarrow {}^3H_4$	5.2	(*16*)
Nd(III)	$^4G_{5/2} \leftarrow {}^4I_{9/2}$	17.3	(*46, 66*)
Pm(III)	$^5G_2, {}^5G_3 \leftarrow {}^5I_4$	18.0	(*67*)
Sm(III)	$^4F_{1/2}, {}^4F_{3/2} \leftarrow {}^6H_{5/2}$	6.4	(*68*)
Eu(III)	$^5D_1 \leftarrow {}^7F_1$	18.7	(*58*)
	$^5D_2 \leftarrow {}^7F_0$	21.5	(*68*)
	$^5D_0 \rightarrow {}^7F_2$	16.3	(*51*)
Gd(III)	$^6P_{5/2}, {}^6P_{7/2} \leftarrow {}^8S_{7/2}$	32.5	—
Dy(III)	$^6F_{11/2} \leftarrow {}^6H_{15/2}$	7.7	(*69*)
Ho(III)	$^5G_6 \leftarrow {}^5I_8$	22.1	(*46, 69*)
	$^3H_6 \leftarrow {}^5I_8$	27.7	(*69, 70*)
Er(III)	$^2H_{11/2} \leftarrow {}^4I_{15/2}$	19.2	(*46, 69*)
	$^4G_{11/2} \leftarrow {}^4I_{15/2}$	26.4	(*46, 69*)
Tm(III)	$^3F_4 \leftarrow {}^3H_6$	5.9	(*33*)
	$^3H_4 \leftarrow {}^3H_6$	12.7	(*69*)
	$^1G_4 \leftarrow {}^3H_6$	21.3	(*33*)

[a] Transitions considered to be potentially hypersensitive from the value of $\Gamma^{(2)}$.
[b] Reference to the observation of hypersensitivity.

these transitions. The $^3P_2 \leftarrow {}^3H_4$ and $^1D_2 \leftarrow {}^3H_4$ of Pr^{3+} are also sensitive to the environment (24, 71) but since these do not obey the selection rules noted above they are not considered hypersensitive in the sense meant in this chapter and will be discussed separately below (Section 6). Two other transitions have also been claimed as hypersensitive: $^4G_{7/2}$, $^2K_{13/2} \leftarrow {}^4I_{9/2}$ of Nd^{3+} (72) and $^6P_{7/2}$, $^4D_{1/2}$, $^4F_{9/2} \leftarrow {}^6H_{5/2}$ of Sm^{3+} (24). Re-examination of the original spectra suggests however that these transitions are not hypersensitive. Some confusion has arisen from the practice of measuring extinction coefficients and not oscillator strengths. A large number of publications have been concerned to some extent with hypersensitive transitions and quite a lot is now known about their behaviour. A summary of this will now be given.

1. The intensity of a hypersensitive transition is zero (except for vibronic intensity) when the lanthanide ion is at a centre of symmetry. This has been established for three compounds of Eu^{3+} (59) and for most of the lanthanides doped into S_6 site of Y_2O_3 (73). Most recently this effect is found (111) in the spectrum of $Cs_2NaEuCl_6$ where pure electronic lines are missing from the $^5D_2 \leftarrow {}^7F_0$ transition. In exhibiting this behaviour the hypersensitive transitions behave exactly like the nonhypersensitive ones.

2. Very small deviations from inversion symmetry can cause the hypersensitive transitions to have quite large intensity while leaving the intensities of the other transitions virtually unaltered. Again this has been demonstrated explicitly (59) for several compounds of Eu^{3+} and for Er^{3+} doped in YCl_3 (74) where the site symmetry is very slightly distorted from O_h. The spectra of the lanthanide hexahalides (52, 53) and hexaisothiocyanates (55) have been interpreted in this light.

3. The intensity of a hypersensitive transition can be up to 200 times greater (33, 34) than that of the corresponding aquo ion transition, depending upon the particular complex, and in only three cases has the intensity been reported (25, 35, 52) as less than that of the aquo ion. The intensities of the other transitions in complexed species are rarely larger than those of the aquo ions and are frequently less.

4. The intensity order with changing ligand for a given hypersensitive transitive transition and constant structure appears to be (33, 34, 35, 52, 53):

$$I^- > Br^- > Cl^- > H_2O > F^-$$

For a series of tris β-diketonate complexes the aryl substituted ligands give rise to larger intensities than do the alkyl substituted ones (28, 29).

5. Hypersensitivity has been shown to be proportional to the nephelauxetic ratio (75, 76) (and so possibly to co-valency), although in certain cases this relationship does not seem to hold (24).

6. Hypersensitivity is frequency dependent (26) in that for a series of Ho^{3+} and Er^{3+} complexes the Ω_2 parameter obtained from the higher energy hypersensitive transition is larger than that obtained from the lower energy one.

7. Recent work on a series of complexes of Nd^{3+} and Ho^{3+} with weakly co-ordinating unidentate ligands suggests that a correlation exists between the intensity of the hypersensitive transitions and the pKa of the ligand (77).

Hypersensitivity has provoked more interest than any other aspect of lanthanide intensities and a number of theories both qualitative and quantitative have been put forward to explain it.

5.2. An Explanation within the Framework of the Judd-Ofelt Theory

If it is assumed that the radial integrals $\langle 4f | r^n | 4f \rangle$ and $\langle 4f | r^n | 5d \rangle$ are not affected to any large extent by the environment of the lanthanide ion, the only way in which Ω_2 as defined by Eq. (13b) can be altered without a corresponding change in at least Ω_4 is through the A_{1p} crystal field parameters (which occur only in the expression for Ω_2). In his original paper (6) and in subsequent work (46) Judd decided that A_{1p} crystal field parameters were not admissable in the expression for Ω_2 since their presence would imply a non-vanishing electric field at the lanthanide ion nucleus and hence that the ion was not in equilibrium with its surroundings which is a basic requirement in a static perturbation model. In a later paper (78), however, Judd re-examined the possibility of including A_{1p} in the expression for Ω_2 and decided that since the electrons of the lanthanide ion can also produce (under the influence of the perturbation) a non-zero electric field at the nucleus which exactly cancels out that produced by the crystal field, A_{1p} could be included in the lattice sum without going against the spirit of the model. Other authors (79, 80) have given arguments both theoretical and practical for including A_{1p} in the crystal field expansion. Only certain point groups permit A_{1p} crystal field parameters. They are C_s, C_1, C_2, C_3, C_4, C_6, C_{2v}, C_{3v}, C_{4v} and C_{6v} and so only in complexes belonging to these point groups would hypersensitivity be predicted.

In order to test this theory we must know the accurate site symmetry of a lanthanide ion in a complex or crystal host and observe whether or not hypersensitivity is exhibited. A group of examples of ions doped into hosts where the nearest neighbours are all oxide ions appear to conform to the theory (Table 1). Ω_2 for Nd^{3+} in $YAlO_3 (D_{3h})$ and Yttrium aluminium garnet (D_{2d} or D_2) is either similar to or less than that for the aquo ion while in $Y_2O_3 (C_2)$ it is very much greater than the aquo ion value. The $Nd(NO_3)_9$ ion (C_3) is also very hypersensitive (81) while Ω_2 for both Eu^{3+} and Tm^{3+} ethylsulphates (C_{3h}) are lower than the values for the corresponding aquo ions.

There are, however, exceptions to the theory, the most striking being the gaseous trihalides LnX_3 (X = Br, and I). $NdBr_3$ and NdI_3 have been shown (82) to have D_{3h} symmetry in the gas phase by electron diffraction techniques and yet have the largest Ω_2 values known for any complexes of Nd(III). The 'octahedral' hexahalide ions have, in some cases, hypersensitive transitions which are stronger than in the aquo ions. Although presumably distorted to some extent it is unlikely that the symmetry would be so low as to contain Y_{1m} harmonics. Finally Pappallardo has noted (83) that the Stark splitting pattern of the very hypersensitive transition of Holmium tris(methylcyclopentadienyl) is not compatible with any of the symmetries containing A_{1p} in their crystal field expansion.

It is obvious then that although this theory can probably account for some aspects of hypersensitivity it cannot be the whole explanation.

5.3. Vibronic Mechanism

An assessment of the importance of a vibronic versus a static mechanism for the Ω_λ parameters in general has been given above and it can only be stressed again that all the direct experimental evidence points to a vibronic mechanism being unimportant at least for molecular complexes. It has been shown for several complexes that the intensity of the $^5D_2 \leftarrow {}^7F_0$ transition of Eu(III) *when exhibiting hypersensitivity* is largely static (*48, 51, 60*) and that when the complex possesses a centre of inversion it is very nearly zero (*59*). Little vibronic structure has been found in the 4 K spectra of the hypersensitive bands of $[(C_6H_5)_3PH]_3NdCl_6$ (*54*), Tm(methylcyclopentadienyl)$_3$ (*83*) and Er(cyclopentadienyl)$_3$ THF (*84*). *Krupke* has interpreted (*35*) the variation of the Ω_2 parameters of Ln^{3+} in the Y_2O_3 host as implying that they contain little contribution from a vibronic perturbation in contrast to Ω_4 and Ω_6 and a similar conclusion was reached (*42*) for the Ω_2 parameter of Tm ethylsulphate.

Jørgensen and *Judd* (*46*) considered the possibility of vibrations which could contribute to Ω_2 but not to Ω_4 or Ω_6 (the analogue of A_{1p} in the static model) and produced the equation

$$\Omega_2(\text{VIB}) \approx \left[\frac{h\, a_0^4}{4\pi^2\, m\, \chi}\right] \frac{N\,(\varrho'')^2}{(\varrho')^6}\, (2J+1)^{-1} \times 10^{-16}\, \text{cm}^2 \qquad (19)$$

where a_0 is the Bohr radius, ϱ' is the radius of the vibrating complex and ϱ'' the amplitude of vibration. N is a dimensionless geometric factor depending on the structure of the complex and is about 10.

Taking the very reasonable values of 4Å for ϱ' and 0.08 Å for ϱ'' they calculated a value of Ω_2 (vibronic) which was about two orders of magnitude too small. Eq. (19) has been used by *Gruen et al.* (*33*) to calculate values of Ω_2 for gaseous $NdBr_3$ and NdI_3 which are of the same order of magnitude as the experimental ones. Their choice of values for ϱ' and ϱ'' has, however, been criticised (*60b*) and in particular, because the Nd-I bond is longer than that of Nd-Br (*82*), application of this vibronic formula must lead to the conclusion that Ω_2 for NdI_3 is smaller than that for $NdBr_3$ which is contrary to experiment.

The presence of strong hypersensitive transitions in the spectra of the octa-hedral' hexahalides has been claimed (*75*) as evidence that Ω_2 for these complexes is largely vibronic, but it seems more likely (based on the 4 K crystal spectrum of $[(C_6H_5)_3PH]_3NdCl_6$) that a small distortion from pure octahedral symmetry is present. This conclusion was also reached by *Ryan* and *Jørgensen* (*52*) on the basis that the pure electronic lines in the spectrum of the Pr(III) complexes has significant intensity. A vibronic mechanism has also been invoked (*85*) to rationalise the changes in the spectrum of an aqueous solution of Nd^{3+} nitrate when it is heated to 620 K. Again no concrete evidence is presented and the spectra suggest that three different hydrates (with presumably different symmetries) are formed at various temperatures.

5.4. The Inclusion of Covalency

Both static and vibronic intensity can be increased if a significant amount of covalency is postulated. The nephelauxetic ratio has long been regarded (*86*) as a measure of covalency and the fact that the intensity of the hypersensitive band of Nd^{3+} can be correlated with this ratio has been considered (75) an indication that covalency is important. From Eq. (10) it can be seen that the inclusion of covalency can affect both $\langle 4f |r^t| nl \rangle$ and $\langle f \| C^{(t)} \| l \rangle$. The former cannot be responsible for hypersensitivity since any perturbation of the $4f$ radial function would affect all the Ω_λ parameters. It has, however, been argued (75) that the angular part of the $4f$ function can also be modified by covalency and that this can preferentially affect Ω_2.

The ligand wave functions may be written (87) as a sum of (renormalised) radial functions multiplied by spherical harmonics with origin at the central metal ion.

$$\Phi\text{ligands} = \sum_k \alpha_k \sum_m Y_{km} \qquad \text{with } k \text{ odd.}$$

The metal function is now written as a sum of the original $4f$ functions and the ligand functions weighted by mixing coefficients related to the nephelauxetic ratio.

$$\Phi\text{metal} = (1 - b)^{\frac{1}{2}} |4f\rangle - b^{\frac{1}{2}} |\Phi\text{ligands}\rangle .$$

The matrix element $\langle f \| C^{(t)} | l\rangle$ of Eq. (10) is now replaced by

$$(1 - b)^{\frac{1}{2}} \langle f \| C^{(t)} \| l \rangle - b^{\frac{1}{2}} \langle \Phi\text{ligands} \| C^{(t)} \| l) .$$

The relevant term which can affect the value of Ω_λ is then

$$- b^{\frac{1}{2}} \sum_k \alpha_k \langle Y_{km} \| C^{(t)} \| l \rangle .$$

The authors then argue (75) that if $4f - 5d$ mixing only occurs then Ω_2 and to a lesser extent Ω_4 will be affected while Ω_6 should be unaffected by the environment. It seems clear, however, that if $4f - ng$ mixing is included then Ω_6 ought also to be sensitive according to this mechanism. There is little evidence that transitions involving Ω_4 are hypersensitive, for example the intensity of the $(^5G, ^2G)_5 \leftarrow {}^5I_8$ transition of Ho^{3+} is dominated by $\Gamma^{(4)}$ and seems little affected by the environment.

5.5. Pure Quadrupole Radiation

As discussed in Section 2.8 $f \leftrightarrow f$ transitions can occur by the absorption of quadrupole (and hexadecapole and 64-pole) radiation. The expression for the electric quadrupole oscillator strength has already been given Eq. (18). The selection rules on quadrupole transitions are those on $\Gamma^{(2)}$ i.e. $|\Delta J| \leqslant 2$, $|\Delta L| \leqslant 2$ and $\Delta S = 0$ and so quadrupole radiation is presumably absorbed during hypersensitive transitions. The values of P calculated by considering this mechanism are, however, between 10^3 and 10^5 times less than the experimental values. Two further points argue against hypersensitive transitions being electric quadrupole: the fact that in no case where the selection rules on the Stark transitions of a lanthanide spectrum have been determined have quadrupole selection rules been found

109

(19) and the fact that Eq. (18b) implies that Ω_2 should be proportional to σ^2, which is not experimentally found (26).

Jørgensen and *Judd* pointed out (46) that an enhancement of the quadrupole oscillator strength could be obtained if the $\langle 4f|r^2|4f\rangle$ radial integral was modified by co-valency. As in the previous section the $4f$ radial wavefunction is replaced by

$$(1-b)^{\frac{1}{2}}|4f\rangle - b^{\frac{1}{2}}|\Phi\text{ligands}\rangle$$

Taking reasonable values of b, the oscillator strength could be increased by a factor of about 100 but it is still left smaller than the experimental value. It is perhaps interesting in this connection to note that Pappalardo had reported (84) a correlation between the (large) oscillator strengths of the hypersensitive bands of Nd(III) and Er(III) cyclopentadienyl complexes and a large splitting of the $J=3/2$ levels, which is directly proportional to $\langle 4f|r^2|4f\rangle$. The problems of selection rules and frequency dependence, however, remain.

5.6. Inhomogeneous Dielectric

This is the mechanism finally proposed (46) by *Jørgensen* and *Judd* to account for hypersensitivity. They noted that the electric quadrupole operator which is normally written

$$e(^2/_3)^{\frac{1}{2}} \sum_i r_i^2 (C_q^{(2)})_i$$

should more properly be

$$e(^2/_3)^{\frac{1}{2}} \sum_i r_i^2 (C_q^{(2)})_i ((\nabla E)_p^{(2)})_i$$

where

$$(\nabla E)_p^{(2)}$$

represents the variation of the electric field across the ion and is normally considered to be insignificant (*i.e.* is taken as unity). It was argued, however, that were the electric field variation not negligible, due to the presence around the ion of dipoles induced in the surroundings by the radiation field, then the quadrupole oscillator strength could be greatly enhanced. Taking as a model a cubic array of ions in which one ion had a polarisability twice that of the others they found that the oscillator strength could be enhanced thus:

$$\Omega_2 \text{ (inhomogenous dielectric)} = \Omega_2 \text{ (quadrupole)} \cdot \frac{225}{16\,\pi^4} \cdot \frac{(n^2-1)^2}{n^2(n^2+2)^2} \cdot \frac{a^6}{R^8\sigma^2} \quad (20)$$

where n is the refractive index of the medium and R is the distance from the lanthanide ion to the unique and a to the non-unique surrounding ions. The expression for Ω_2 is now frequency independent, removing one of the major obstacles to a quadrupole mechanism, although it would appear that transitions between Stark components should still obey quadrupole selection rules.

With $n = 1.5$ and $R = a$ the oscillator strength of the $^4G_{5/2} \leftarrow {}^4I_{9/2}$ transition of Nd^{3+} was calculated and the value obtained was only thirty time less than experiment compared to about 10^4 for a pure quadrupole mechanism. The calculated value could now be increased to match the experimental one by the inclusion of covalency.

This theory has been criticised by several authors $(24, 33, 75, 78, 88, 89)$; in many cases without justification. The name 'inhomogeneous dielectric' has led to the mistaken belief $(24, 88)$ that it predicts that hypersensitivity cannot occur in the solid state. Any system in which the lanthanide ion is not at a centre of inversion can, according to the model, give rise to hypersensitivity. One paper (89) discounts the theory on the basis of the observation that the oscillator strength of the hypersensitive bands of some β-diketone complexes did not vary regularly with the polarisability of the *solvent* with no evidence that the solvent was co-ordinating. Finally it has been discounted (33) because of the incorrect conclusion that it could not account for the hypersensitivity of the gaseous LnX_3 (which have become test cases for theories of hypersensitivity). The authors (33) in this case evaluated Eq. (20), using the value of 1.001 for the refractive index. This is clearly inappropriate since it is the bulk refractive index of the gas, and the refractive index intended in the relationship is an approximation to the polarisability of the ligands, or solvent molecules for a solvated Ln^{3+} ion. We feel that the inhomogeneous dielectric theory has not been disproved but because the authors' have not derived a formula which can be applied to a real system it is not clear what the degree of enhancement of quadrupole intensity would be in any particular case.

5.7. Dynamic Coupling Mechanism

This is the most recent mechanism proposed $(90, 91)$ to account for hypersensitivity. In the Judd-Ofelt theory the ligands surrounding the lanthanide ion are considered only inasmuch as their ground state field produces the perturbation required to mix excited configurations into the $4f^N$ configurations. However the ligand wave-functions are also perturbed by the metal ion. If we write the metal ground and excited functions as $|M_0\rangle$ and $|M_a\rangle$ and those of the ligand as $|L_0\rangle$ and $|L_b\rangle$ and represent the total system of (lanthanide ion and ligand) by simple product functions, $i.e.$ $|M_0L_0\rangle$, we may write (92) the first order perturbed ground and excited states $|A\rangle$ and $|B\rangle$ as:

$$|A\rangle = |M_0L_0\rangle - \sum_b (E_a + E_b)^{-1} \langle M_a L_b |V| M_0 L_0 \rangle |M_a L_b\rangle$$

$$|B\rangle = |M_a L_0\rangle + \sum_b (E_a - E_b)^{-1} \langle M_0 L_b |V| M_a L_0 \rangle |M_0 L_b\rangle$$

where the energies of the $|M_a\rangle \leftarrow |M_0\rangle$ and $|L_b\rangle \leftarrow |L_0\rangle$ transitions are given by E_a and E_b and the summation runs over all transitions of the ligand which are electric dipole allowed.

The $f \leftrightarrow f$ transition thus aquires a first order electric dipole transition moment which is given by:

$$e\langle A |\boldsymbol{D}_q^{(1)}|B\rangle = e \sum_b 2E_b(E_b^2 - E_a^2)^{-1}\langle M_0M_a|\boldsymbol{V}|L_0L_b\rangle\langle L_0|\boldsymbol{D}_q^{(1)}|L_b\rangle \qquad (21)$$

The electric dipole moments $\langle L_0|\boldsymbol{D}_p^{(1)}|L_b\rangle$ located on the ligand are correlated coulombically with the charge distribution at the lanthanide ion caused by the $f \leftrightarrow f$ transition. This correlation is given by the matrix element $\langle M_0M_a|\boldsymbol{V}|L_0L_b\rangle$ where the potential $\boldsymbol{V}$ represents the coulombic interaction of the charge distributions on metal and ligand. This matrix element can be evaluated by expanding the transitional charge distributions M_0M_a and L_0L_b as multipole series centred on their respective co-ordinate origins. The leading multipole of each is retained; an electric dipole for the ligand charge distribution and an electric quadrupole for the metal. The perturbation matrix element now becomes (93):

$$\langle M_0M_a|\boldsymbol{V}|L_0L_b\rangle =$$
$$- \frac{e^2}{R^4} \sum_q \sum_m (-1)^{q+m} [B^{q,m}]^{\ddagger} \, C^{(3)}_{-q-m}\langle M_0|\boldsymbol{D}_m^{(2)}|M_a\rangle\langle L_0|\boldsymbol{D}_q^{(1)}|L_b\rangle \qquad (22)$$

where R is the metal-ligand bond distance and $B^{q,m}$ is a numerical factor depending on the particular components q and m which interact.

If we restrict our consideration to N perturbers (ligands) L each at a common distance R from the lanthanide ion we have

$$e\langle A |\boldsymbol{D}_q^{(1)}|B\rangle = R_e^{-4} \sum_{L=1}^{N} \sum_{q,m} (-1)^{m+q}[B^{q,m}]^{\ddagger} \, C^{(3)}_{-q-m}\langle M_0|\boldsymbol{D}_m^{(2)}|M_a\rangle$$
$$\left\{ \sum_b \frac{2E_b\, e^2 \langle L_0|\boldsymbol{D}_q^{(1)}|L_b\rangle^2}{(E_b^2 - E_a^2)} \right\} \qquad (23)$$

The content of the curly brackets is just the diagonal element of the polarisability tensor α_{qq} at the frequency of the transition, and, if we restrict ourselves to isotropic ligands, may be equated to the average polarisability α. Finally if we revert to the more normal S, L, J, M notation (i.e. $|M_0\rangle$ goes to $|4f\alpha[SL]JM\rangle$] and sum over the q components of the dipole and m components of the quadrupole operators and sum over M and M' as is appropriate for isotropic solution the oscillator strength becomes:

$$P(\text{dynamic}) = \chi\left[\frac{8\pi^2 m c}{3h}\right] \sigma \,\Omega_2(\text{dynamic})\langle f^N\alpha[SL]J\|U^{(2)}\|f^N\alpha'[S'L']J'\rangle^2$$
$$\times (2J+1)^{-1} \qquad (24a)$$

where

$$\Omega_2(\text{dynamic}) = \left(\frac{2}{15}\right)\left(\frac{\bar{\alpha}^2}{R^8}\right)\langle f\|C^{(2)}\|f\rangle^2 \langle 4f|r^2|4f\rangle^2 \sum_{q,m} \left\{ [B^{q,m}] \left| \sum_{L=1}^{N} C^{(3)}_{-q-m} \right|^2 \right\}$$
$$(24b)$$

The dynamic coupling mechanism, then, gives a contribution to Ω_2 and only Ω_2 in addition to that from the Judd-Ofelt mechanism and so is clearly a possible explanation of hypersensitivity. It predicts that hypersensitivity should be exhibited if the point group of the lanthanide contains the Y_{3m} spherical harmonic in the expansion of its point potential. The appropriate point groups are all those containing the Y_{1m} harmonic as discussed in Section 5.2 above (C_S, C_1, C_2, C_3, C_4, C_6, C_{2v}, C_{3v}, C_{4v} and C_{6v}) and in addition C_5, C_7, C_8, C_{5v}, $C_{\infty v}$, C_{3h}, D_2, D_3, D_4, D_5, D_6, D_{2d}, D_{3h}, S_4, T and T_d. Thus we have the possibility of hypersensitivity for the D_{3h} trihalides. The theory has been applied quantitatively to NdBr$_3$ and NdI$_3$. Depending on the values of radial integrals and polarisabilities used Ω_2 for NdBr$_3$ is calculated in the range $100-170 \times 10^{-20}$ cm^2 and for NdI$_3$ between $140-230 \times 10^{-20}$ cm^2, so not only are the calculated values close to experiment but Ω_2 for the tri-iodide is predicted to be larger than that for the tribromide as is found experimentally. Various properties of hypersensitivity can be accounted for in terms of the polarisability which occurs in Eq. (24 b). For example the order of polarisabilities is

$$I^- > Br^- > Cl^- > H_2O > F^-$$

and aryl $>$ alkyl which is the order of hypersensitive intensity; polarisability is a function of frequency and of covalency and can probably be correlated with the pKa of the corresponding acid. The extreme sensitivity to environment can be rationalised both by the polarisability and by the presence of the R^{-8} factor in Eq. (24 b).

The principal difference between the dynamic coupling and inhomogeneous dielectric theories is that in the latter dipoles are induced in the ligands by the radiation field and are to some extent random and mutually cancelling while in the former the dipoles are induced by the charge distribution caused by the $f \leftrightarrow f$ transition and so are correlated to give maximum effect. Although the dynamic coupling theory must of course be subjected to more test it does seem at the present to be the most able to account both qualitatively and quantitatively for hypersensitivity.

5.8. Conclusion

It is felt on the basis of the evidence reviewed above that the *Judd*, the inhomogeneous dielectric and the dynamic coupling theories may all be considered to some extent to account for hypersensitivity. Unfortunately only in the case of the last named have quantitative calculations been carried out and it is not yet certain what the relative quantitative contributions to the phenomenon are in any real case.

6. The Exceptional Behaviour of Praseodymium (III)

The phenomenological application of the Judd-Ofelt theory works less well for complexes of Pr(III) than for those of the other lanthanides. This is evidenced both by the much larger RMS values obtained and by the difficulty experienced

(17) when trying to fit both the $^3F_3$, $^3F_4 \leftarrow {}^3H_4$ and $^3P_{2,1,0}{}^1I_6 \leftarrow {}^3H_4$ transition groups with the same phenomenological parameters. The Judd-Ofelt parameters for a number of compounds of Pr(III) are collected in Table 3. The Ω_2 parameters are, in many of the cases, meaningless since only the $^3F_2 \leftarrow {}^3H_4$ transition has a significant $\Gamma^{(2)}$ matrix element and it has not been included in the data set of any of the compounds measured in aqueous solution. The parameters in the first row of Table 2 have been obtained by extrapolation from those for the other lanthanide aquo ions. A comparison of the values of Ω_λ for at least all the aqueous solutions with the extrapolated values shows that Ω_6 is always considerably larger than would be expected if the smooth variation of Ω_6 with number of f electrons were continued from Nd^{3+} to Pr^{3+}. Table 4 lists the oscillator strengths

Table 3. Judd-Ofelt parameters for compounds of Pr(III)

$\Omega_\lambda (\times 10^{20})$ cm^2	2	4	6	RMS ($\times 10^6$)	Ref.
Extrapolated from Ω_λ of the other $L_n{}^{3+}$ aquo ions	1.2	5.1	9.0		
Pr^{3+} aquo (excluding $^3F_{3,4} \leftarrow {}^3H_4$)	32.8 ± 72	5.8 ± 2.5	32.1 ± 3.0	1.3	(17)
Pr^{3+} aquo (including $^3F_{3,4} \leftarrow {}^3H_4$)	-241.5 ± 115	13.5 ± 5.3	25.1 ± 7.1	3.7	(17)
Pr (α-picolinate)	46 ± 72	7.0 ± 2.5	39.5 ± 3.0	1.3	(23)
$[PrW_{10}O_{35}]^{7-}$	-27.2 ± 60	5.1 ± 2.1	37.3 ± 2.5	1.1	(24)
$[Pr(PrW_{11}O_{39})_2]^{11-}$	-79.9 ± 180	7.8 ± 6.2	72.5 ± 7.5	3.3	(24)
$[Pr(P_2W_{17}O_{61})_2]^{17-}$	13.6 ± 102	5.9 ± 3.5	67.1 ± 4.2	1.8	(24)
Pr(NO$_3$)$_3$ in ethyl acetate	1.8 ± 10.3	8.5 ± 3.1	13.7 ± 4.1	3.7	(16)
Pr(NO$_3$)$_3$ in LiNO$_3$/KNO$_3$	8.7 ± 0.4	6.4 ± 1.1	6.5 ± 1.5	1.4	(16)
Pr^{3+} in Y$_2$O$_3$	17.2 ± 0.2	19.8 ± 0.1	4.9 ± 0.1	0.1	(35)
Pr^{3+} in LaF$_3$	0.12 ± 0.9	1.8 ± 0.8	4.8 ± 0.5	1.3	(35)

Table 4. Observed and calculated oscillator strengths ($\times 10^6$) of Pr(III) transitions

Transition	$P_{observed}$	$P_{calculated}{}^{a)}$	$P_{calculated}{}^{b)}$	$P_{calculated}{}^{c)}$
$^3P_2 \leftarrow {}^3H_4$	14.96	4.5	14.6	12.5
$^3P_1, {}^1I_6$	7.54	4.1	7.3	4.1
$^3P_0$	2.51	2.6	3.0	6.9
$^1D_2$	3.03	1.3	4.5	2.2
$^1G_4$	0.30	0.4	1.3	0.7
$^3F_3, {}^3F_4$	12.58	12.1	41.1	14.1

a) From Ω_λ extrapolated from Ω_λ for the other lanthanide(III) aquo ions.
b) By solving Eq. (13a) excluding the $^3F_{3,4} \leftarrow {}^3H_4$ transitions.
c) By solving Eq. (13a) including the $^3F_{3,4} \leftarrow {}^3H_4$ transitions.

of the various transitions of the Pr^{3+} aquo ion along with the oscillator strengths calculated a) from the extrapolated values of Ω_λ b) from the values obtained by *Carnall et al.* *(17)* by excluding and c) including the $^3F_3$, $^3F_4 \leftarrow {}^3H_4$ transitions. The extrapolated parameters reproduce quite well the intensities of the $^3P_0$, $^1G_4$ and $^3F_{3,4}$ transitions while underestimating those of the $^3P_{2,1}$, $^1I_6$ and $^1D_2$ groups. Conversely if those transitions are fitted (calculation b) then $^1G_4$, $^3F_{3,4}$ and to a lesser extent $^3P_0$ are overestimated. Attempting to fit all transitions with the same parameters (calculation c) leads to poor agreement in almost all cases. All the non-aqueous complexes listed in Table 2 have been treated by method (c) and good agreement between calculated and experimental oscillator strengths is found only for Pr^{3+} in LaF_3 *(35)*.

It is possibly significant that the $^3P_2 \leftarrow {}^3H_4$ and $^1D_2 \leftarrow {}^3H_4$ transitions have been found *(71)* to exhibit hypersensitivity (although this word should be really be reserved for transitions obeying quadrupole selection rules as discussed in Section 5). Figure 5 shows the spectrum of an aqueous solution of $K_{11}[Pr(PW_{11}O_{39})_2]$ and of Pr^{3+} perchlorate for comparison. The same enhancement of intensity

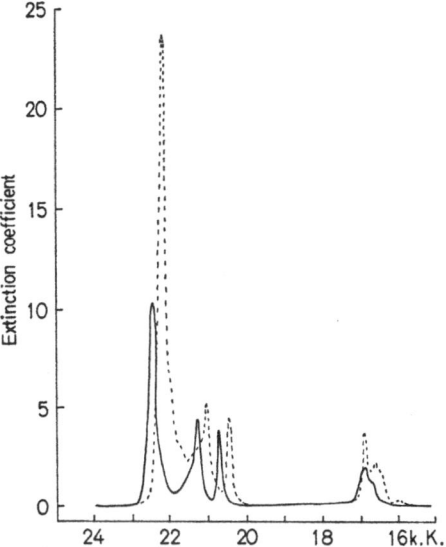

Fig. 5. The aqueous solution spectra of $[Pr(PW_{11}O_{39})_2]^{11-}$ (broken line) and Pr^{3+} perchlorate (full line) [from (Ref. *112*)]

obtains in several other heteropolycomplexes *(24, 71)* and in some β-diketone complexes of Pr(III) *(27)*. The $^3P_2 \leftarrow {}^3H_4$ transition has about half the uqao ion intensity in the KNO_3-$LiNO_3$ melt *(16)* and the entire $^3P_{2,1,0}$ $^1I_6 \leftarrow {}^3H_4$ and $^1D_2 \leftarrow {}^3H_4$ groups of transitions are absent *(33)* from the spectrum of $PrBr_3$ and PrI_3, although the $^3F_{3,4} \leftarrow {}^3H_4$ transition is present with much the same intensity as in the aquo ion. It has been claimed *(94)* that the $^1I_6 \leftarrow {}^3H_4$ transition is hypersensitive.

115

The theory commonly (*17*, *95*) put forward to explain the anomalous behaviour of Pr^{3+} is that the approximation made in the Judd-Ofelt theory that the perturbing configurations are degenerate and of much higher energy than the $f \leftrightarrow f$ transition being considered is not valid. The first $4f \rightarrow 5d$ absorption of Pr^{3+} in a CaF_2 host is located at about 45,000 cm^{-1} while that of any of the other lanthanides (except Tb^{3+}) is at least 10,000 cm^{-1} to higher energy. The Tb^{3+} ion has a $4f \rightarrow 5d$ transition at essentially the same energy as that of Pr^{3+} but arguments have been advanced (*17*) as to why the breakdown of the approximation is less serious for Tb^{3+}.

The breakdown of this approximation for Pr^{3+} affects two aspects of the Judd-Ofelt theory. Firstly it makes the closure argument rather more dubious and secondly it makes the cancellation of the 'odd' parts of the intensity expression less exact. This may result in a significant contribution from

$$\sum_\lambda \Omega_\lambda \langle f^N[SL]J\|U^{(\lambda)}\|f^N[S'L']J'\rangle^2$$

with λ *odd*. In fact for the $^3P_{2,1,0} \leftarrow {}^3H_4$ multiplet only the $\lambda = 5$ reduced matrix elements are non-zero and have values of ~ 0.283 for $^3P_2 \leftarrow {}^3H_4$, ~ 0.502 for $^3P_1 \leftarrow {}^3H_4$ and zero for $^3P_0 \leftarrow {}^3H_4$. It is possible, then, that Ω_5 can contribute to the intensity of at least the $^3P_2 \leftarrow {}^3H_4$ and $^3P_1 \leftarrow {}^3H_4$ transitions. This cannot be the whole explanation, however, since the greatest discrepancy between the observed and extrapolated intensity is for the $^3P_2 \leftarrow {}^3H_4$ transition while the $\Gamma'^{(5)}$ reduced matrix element for $^3P_1 \leftarrow {}^3H_4$ is larger. It is also difficult to explain the observed hypersensitivity by this method. The exceptional behaviour of Pr(III) and in particular its hypersensitivity remain one of the major problems in lanthanide intensity theory.

7. The $^5D_0 \leftrightarrow {}^7F_0$ Transitions of Europium (III)

The $^5D_0 \leftrightarrow {}^7F_0$, $^5D_0 \rightarrow {}^7F_3$, and $^5D_0 \rightarrow {}^7F_5$ and $^5D_3 \leftarrow {}^7F_0$ transitions of Eu(III) are all strictly forbidden within the framework of the first order perturbation treatment of the Judd-Ofelt theory. All three do occur, usually with extremely low intensity, in the spectra of certain compounds and this is usually explained as due to 'J-mixing' *i.e.* the wave functions of $J \neq 0$ states are mixed into the $J = 0$ state by the *even* parity terms of the crystal field. For example for the $^5D_0 \leftrightarrow {}^7F_0$ transition the total (second order) perturbation expression would be

$$\sum_J \sum_K \langle f^N[^7F]0|V^{\text{even}}|f^N[^7F]J\rangle \langle f^N[^7F]J|V^{\text{odd}}|\psi''\rangle \langle \psi''|D_q^{(1)}|f^N[^5D]0\rangle$$

The $J = 0 \rightarrow J = 3$ or 5 transitions are usually very weak, having oscillator strengths an order of magnitude less than the magnetic dipole allowed $^5D_1 \leftarrow {}^7F_0$ or $^5D_0 \rightarrow {}^7F_1$ transitions and can be easily accounted for by the 'J-mixing' mechanism. The $J = 0 \rightarrow J = 0$ transitions can, however, be quite intense. In Eu(III) doped into Sr_2TiO_4 for example the $^5D_0 \rightarrow {}^7F_0$ intensity is 1.65 times the $^5D_0 \rightarrow {}^7F_1$

intensity (*79, 96*) and other cases are reported in the same paper where the intensity ratio between these two transitions is 0.72 and 0.25. Two attempts to calculate the intensity of a $J = 0 \rightarrow J = 0$ transition have been published; in the case of EuAlO$_3$ (C_{2v}) it was concluded that J-mixing could (*97*) account for the observed intensity while for several alkali europium tungstate compounds (*e.g.* NaEu(WO$_4$)$_2$, S_4) it was decided (*48*) that it could not.

It was noted (*96*) that a $J = 0 \rightarrow J = 0$ transition had large intensity only when the site symmetry of the Eu^{3+} ion was one of those which had A_{1p} terms in its crystal field expansion. For example the transition has not been observed in Eu:Ba$_2$GdNbO$_6$ (O_h), Eu: Gd$_2$Ti$_2$O$_7$ (D_{3h}), Eu^{3+} ethylsulphate (C_{3h}) or Eu: YVO$_3$ (D_{2d}) while it occurs very strongly in Eu: Sr$_2$TiO$_4$ (C_{4v}) Eu: Gd$_2$O$_3$ (S_6 and C_{2v}) and Eu:Ba$_3$Gd$_2$WO$_9$ (C_{3v}). The authors also found (*96*) that the largest $J = 0 \rightarrow J = 0$ intensity occurred in those compounds where a large value of the A_{1p} crystal field parameters would be expected from the crystal structure. They then argued that $J = 0 \rightarrow J = 0$ transitions could have finite intensity if A_{1p} crystal field terms were included in the expansion through the equation.

$$ P = \chi \left[\frac{8 \pi^2 m c}{3 h} \right] \sigma \Omega_0 \langle f^N \alpha [S L] 0 \| U^{(0)} \| f^N \alpha'[S'L'] 0 \rangle^2 $$

where

$$ \Omega_0 = \sum_p A_{1p} \, \Xi \, (1,1) $$

Wybourne has however argued (*98*) that since $U^{(0)}$ is a scalar and the functions $|f^N \alpha[S L] J\rangle$ are orthonormal the reduced matrix element is zero and so that no intensity can come from this mechanism. This argument depends upon the closure procedure being valid and this has now been challenged (*99*). It is rather important to have this controversy resolved since potentially the presence or absence of the $J = 0 \leftrightarrow J = 0$ line in a Eu(III) spectra could be diagnostic of the importance of the A_{1p} terms of the crystal field expansion of the Eu^{3+} ion.

8. Structural Implications

The splitting by the ligand field of J-states into Stark components and the selection rules between these components have long been used to determine the site symmetry of lanthanide ions and, in favourable cases, their co-ordination geometry. The fluorescence of Eu(III) has probably been used in this way to a greater extent than the spectrum of any other lanthanide ion. The nephelauxetic ratio can also give some structural information in that one can often distinguish between co-ordination by, say, nitrogen, oxygen or sulphur by means of it.

It is natural, then, that attempts should have been made to relate the intensity of $f \leftrightarrow f$ transitions, and particularly of the hypersensitive transitions, to the co-ordination environment of the lanthanide ion. Thus the intensity of the hypersensitive transition of Nd^{3+} has been used (*100*) to help distinguish between inner and outer sphere complexes of that ion and it has been shown empirically (*101*) (with reference to the potential use of Nd^{3+} as a probe for enzyme sites) that one can distinguish between the binding of simple carboxyl and α-aminocarboxyl ligands. Much work on the relationship between the intensity, and more commonly the shape, of the hypersensitive bands of Nd^{3+}, Ho^{3+} and Er^{3+} and the co-ordination of the lanthanide ion has been performed by *Karraker* (*102, 103, 104, 105*). He has suggested (*102*), for example, by comparison of the shapes and intensities of the hypersensitive bands with those of complexes of known co-ordination number that the tris acetylacetonates of these lanthanides each contain one molecule of co-ordinated water. By the same method he has deduced (*103*) the co-ordination numbers of the hydrated lanthanide acetates and has proposed (*104*) that the Nd^{3+} ion is nine co-ordinate in dilute aqueous solution but becomes eight co-ordinate in the presence of large concentrations of Cl^- ion.

Until the mechanism responsible for hypersensitivity is fully elucidated, however, all such correlations must remain empirical and it must be admitted that little meaningful structural information has yet been obtained from the intensities of hypersensitive transitions. Each of the Judd, inhomogeneous dielectric and dynamic coupling theories of hypersensitivity do in principle enable structural assignments to be made since by each can Ω_2 be calculated from a geometric model. In this connection one can cite the compound NdI_4Cs; recently obtained (*106*) in the gas phase. The oscillator strength of the hypersensitive transition of this molecule is (*106*) about half that of NdI_3. The two most likely geometries are a distorted tetrahedron (C_{3v}) or a distorted square plane (C_{2v}). Calculation of the oscillator strength expected for each of these geometries by the dynamic coupling theory suggests that unless the bond lengths are very much larger than in NdI_3 the distorted tetrahedron is ruled out because in this case the oscillator strength is predicted to be greater than that for NdI_3.

One interesting case of structural assignment via the intensities of the non-hypersensitive transitions has been expected. The plot of Ω_6 for the lanthanide aquo ions against that for the $[LnW_{10}O_{35}]^{7-}$ series, which is known to be isostructural (*107*), is linear (*25*) and this has been interpreted as suggesting that the aquo ions also are isostructural (at least from Nd^{3+} to Er^{3+}). Recent conductance measurements on the lanthanide perchlorates support (*108*) this assignment.

9. Concluding Remarks

In the subject of the intensities of forbidden transitions, as in many other subjects in Chemistry and Physics, theory has developed somewhat faster than experiment. It is sad to think that, during the 1960's when so much fine experimental work was done on the energy levels of ions in crystals, in virtually no case were the intensities of the transitions measured. However the situation has, as I hope this review has shown, improved in recent years and although of course no theory can ever be proved, the Judd-Ofelt theory can at least be accepted as a good working hypothesis from which to plan further experiments. Although little use has yet been made of lanthanide intensities in structural predictions this perhaps will come when the mechanisms of hypersensitivity have been more thoroughly elucidated. In this connection the author is informed that Professor *R. W. Schwartz*, of the Louisiana State University is initiating a study of the absorption and magnetic circular dichroism spectra of single crystals of octahedral $Cs_2NaLnCl_6$ which should help our understanding of this problem. The interest generated in lanthanide spectral intensities has perhaps to some extent been responsible for the recent renewal of interest in the intensities of transition metal spectra. The dynamic coupling mechanism is now being applied to the intensities of D_3 complexes of Co(III) *(109)* and to tetrahedral complexes of Co(II) and Ni(II) *(110)*.

To look to the future, the author sees the main directions of research both experimental and theoretical being the explanation of hypersensitivity and of the peculiar spectral behaviour of Pr(III), and the calculation of A_{tp} parameters and so Ω_λ by more realistic methods than the point charge crystal-field model. The author would finally like to mention that a review on lanthanide intensities and in particular hypersensitivity is being prepared by Professors *G. R. Choppin, D. E. Henrie* and *R. L. Fellows*.

Acknowledgements. I would like to thank all those who have been involved with me in research into lanthanide spectra over the last few years: Drs. *T. J. R. Weakley* and *J. C. Barnes* of the University of Dundee and Professor *S. F. Mason* and Messrs. *B. Stewart* and *J. A. Hearson* of King's College. I am very grateful to Imperial Chemical Industries Limited for a Fellowship during which the bulk of this review was written.

10. References

1. *Van Vleck, J. H.:* J. Chem. Phys. *41*, 67 (1937).
2. a) *Woudenberg, J. P. M.:* Physica *9*, 217, 936 (1942);
   b) *Franzen, P., Woudenberg, J. P. M., Gorter, C. J.:* Physica *10*, 693 (1943).;
   c) *Hoogschagen, J., Snoek, A. P., Gorter, C. J.:* Physica *11*, 518 (1943);
   d) *Hoogschagen, J., Scholte, Th., Kruyer, S.:* Physica *11*, 504 (1946);
   e) *Hoogschagen, J.:* Physica *11*, 513 (1946).
3. *Broer, L. J. F., Gorter, C. J., Hoogschagen, J.:* Physica *11*, 231 (1945).
4. *Hoogschagen, J., Gorter, C. J.:* Physica *14*, 197 (1948).
5. *Stewart, D. C.:* Argonne National Laboratory Rep. AND4812 (1952).
6. *Judd, B. R.:* Phys. Rev. *127*, 750 (1962).
7. *Ofelt, G. S.:* J. Chem. Phys. *37*, 511 (1962).
8. a) *Racah, G.:* Phys. Rev. *62*, 438 (1942);
   b) *Racah, G.:* Phys. Rev. *63*, 364 (1943).
9. *Edmonds, A. R.:* Angular momentum in quantum mechanics. Princeton: Princeton University Press 1960.
10. *Judd, B. R.:* Operator techniques in atomic spectroscopy. New York: McGrawHill Book Company 1963.
11. *Wybourne, B. G.:* Spectroscopic properties of rare earths. New York: Interscience (John Wiley) 1965.
12. *Shore, B. W., Menzel, D. H.:* Principles of atomic spectra. New York: John Wiley and Sons, Inc. 1968.
13. *Rotenberg, M., Bivins, R., Metropolis, N., Wooten, J. K.:* The 3—j and 6—j symbols. Cambridge, Massachusetts: M. I. T. Press 1959.

14. *Nielson, C. W., Koster, G. F.:* Spectroscopic coefficients for p^n, d^n and f^n configurations. Cambridge, Massachusetts: M. I. T. Press 1964.
15. *Griffith, J. S.:* Mol. Phys. *3*, 477 (1960).
16. *Carnall, W. T., Fields, P. R., Wybourne, B. G.:* J. Chem. Phys. *42*, 3797 (1965).
17. *Carnall, W. T., Fields, P. R., Rajnak, K.:* J. Chem. Phys. *49*, 4412 (1968).
18. *Weber, M. J., Matsinger, B. H., Donlan, V. L., Surrat, G. T.:* J. Chem. Phys. *57*, 562 (1972).
19. *Dieke, G. H.:* Spectra and energy levels of rare earth ions in crystals. New York: Interscience 1968.
20. *Detrio, J. A.:* Phys. Rev. B. *4*, 1422 (1971).
21. *Gashurov, G., Sovers, O. J.:* J. Chem. Phys. *50*, 429 (1968).
22. *Carnall, W. T., Fields, P. R., Rajnak, K.:* J. Chem. Phys. *49*, 4424, 4443, 4447 and 4450 (1968).
23. *Bukietynska, K., Choppin, G. R.:* J. Chem. Phys. *52*, 2875 (1970).
24. *Peacock, R. D.:* J. Chem. Soc. A *1971*, 2028.
25. *Peacock, R. D.:* Mol. Phys. *25*, 817 (1973).
26. *Peacock, R. D.:* Chem. Phys. Letters *16*, 590 (1972).
27. *Tandon, S. P., Mehta, P. C.:* J. Chem. Phys. *52*, 4313 (1970).
28. *Tandon, S. P., Mehta, P. C.:* J. Chem. Phys. *53*, 414 (1971).
29. *Isobe, T., Misumi, S.:* Bull. Chem. Soc. Japan *47*, 281 (1974).
30. *Bukietynska, K., Vadura, R.:* Bull. Acad. Polon. Sci. Ser. Sci. Chim. *22*, 139 (1974).
31. *Kazanskaya, N. A.:* Opt. Spectr. (USSR) (English Transl.) *29*, 587 (1970).
32. *Sinha, S. P., Mehta, P. C., Surana, S. S. L.:* Mol. Phys. *23*, 807 (1972).
33. *Gruen, D. M., DeKock, C. W., McBeth, R. L.:* Advan. Chem. Ser. *71*, 102 (1967).
34. *Gruen, D. M., DeKock, C. W.:* J. Chem. Phys. *45*, 455 (1966).
35. *Krupke, W. F.:* Phys. Rev. A *145*, 325 (1966).
36. *Weber, M. J.:* Phys. Rev. *171*, 283 (1968).
37. *Weber, M. J.:* Phys. Rev. *157*, 262 (1967).
38. *Krupke, W. F.:* I. E. E. E. J. Quantum Electron *QE7*, 153 (1971).
39. *Weber, M. J., Varitimos, T. E., Matsinger, B. H.:* Phys. Rev. B. *8*, 47 (1973).
40. *Weber, M. J., Varitimos, T. E.:* J. Appl. Phys. *42*, 4996 (1971).
41. *Axe, J. D., Jr.:* J. Chem. Phys. *39*, 1154 (1963).
42. *Krupke, W. F., Gruber, J. B.:* Phys. Rev. A *139*, 2008 (1965).
43. *Becker, P. J.:* Phys. Status Solidi (B) *43*, 583 (1971).
44. *Mandel, M.:* Appl. Phys. Letters *2*, 197 (1963).
45. *Samelson, H., Brecher, C., Lempicki, A.:* J. Mol. Spectry. *19*, 349 (1966).
46. *Jørgensen, C. K., Judd, B. R.:* Mol. Phys. *8*, 281 (1964).
47. *Richman, I., Satten, R. A., Wong, E. Y.:* J. Chem. Phys. *39*, 1833 (1963).
48. *Yamada, N., Shivnoya, S.:* J. Phys. Soc. Japan *31*, 841 (1971).
49. *Sayre, E. V., Miller, D. G., Freed, S.:* J. Chem. Phys. *26*, 109 (1957).
50. *Haas, Y., Stein, G.:* Chem. Phys. Letters *11*, 143 (1971).
51. *Blanc, J., Ross, D. L.:* J. Chem. Phys. *43*, 1286 (1965).
52. *Ryan, J. L., Jørgensen, C. K.:* J. Phys. Chem. *70*, 2845 (1966).
53. *Ryan, J. L.:* Inorg. Chem. *8*, 2053 (1969).
54. *Gruber, J. B., Menzel, E. R., Ryan, J. L.:* J. Chem. Phys. *51*, 3816 (1969).
55. *Martin, J. L., Thompson, L. C., Radovich, L. J., Glick, M. D.:* J. Am. Chem. Soc. *90*, 4494 (1960).
56. *Burmeister, J. L., Patterson, S. D., Deardorff, E. A.:* Inorg. Chim. Acta *3*, 105 (1969).
57. *Barnes, J. C., Peacock, R. D.:* J. Chem. Soc. A *1971*, 1082.
58. *Peacock, R. D.:* Observations.
59. *Blasse, G., Bril, A., Nieuwport, W. C.:* J. Phys. Chem. Solids *27*, 1587 (1966).
60. a) *Peacock, R. D.:* Chem. Phys. Letters *10*, 134 (1971);
    b) *Peacock, R. D.:* J. C. S. Faraday II *68*, 169 (1972).
61. *Burns, G.:* Phys. Rev. *128*, 2121 (1962).
62. *Rajnak, K.:* J. Chem. Phys. *37*, 2440 (1962).
63. *Newman, D. J.:* Advan. Phys. *20*, 197 (1971).

64. *Selwood, P. W.:* J. Am. Chem. Soc. *52*, 4308 (1930).
65. *Freed, S.:* Rev. Mod. Phys. *14*, 105 (1942).
66. a) *Moeller, T., Brantley, J. C.:* J. Am. Chem. Soc. *72*, 5447 (1950);
    b) *Moeller, T., Jackson, D. E.:* Anal. Chem. *22*, 1393 (1950);
    c) *Moeller, T., Ulrich, W. F.:* J. Inorg. Nucl. Chem. *2*, 164 (1956).
67. *Carnall, W. T.:* J. Phys. Chem. *68*, 2351 (1969).
68. *Carnall, W. T., Gruen, D. M., McBeth, R. L.:* J. Phys. Chem. *66*, 2159, (1962).
69. *Carnall, W. T.:* J. Phys. Chem. *67*, 1206 (1963).
70. *Abrahamer, I., Marcus, Y.:* Israel Atomic Energy Commission Rep. IA-809 (1963).
71. *Peacock, R. D.:* Chem. Phys. Letters *7*, 187 (1970).
72. *Kononenko, K. I., Poluektov, W. S.:* Russ. J. Inorg. Chem. *7*, 965 (1962).
73. *Kisliuk, P., Krupke, W. F., Gruber, J. B.:* J. Chem. Phys. *40*, 3606 (1964).
74. *Rakestraw, J. W., Dieke, G. H.:* J. Chem. Phys. *42*, 873 (1965).
75. *Henrie, D. E., Choppin, G. R.:* J. Chem. Phys. *49*, 477 (1968).
76. *Mehta, P. C., Surana, S. S. L., Bhutra, M. P., Tandon, S. P.:* Spectry. Letters *4*, 181 (1971).
77. *Choppin, G. R., Fellows, R. L.:* J. Co-ord. Chem. in the press; personal communication to the author.
78. *Judd, B. R.:* J. Chem. Phys. *44*, 839 (1966).
79. *Nieuwport, W. C., Blasse, G.:* Solid State Commun. 4, 227 (1966).
80. *Kiss, E. J., Weakliem, H. A.:* Phys. Rev. Letters *15*, 457 (1965).
81. *Carnall, W. T., Siegel, S., Ferraro, J. R., Gebert, E.:* Inorg. Chem. *12*, 560 (1973).
82. a) *Akishin, P. A., Naumovard, V. A., Tatevskii, V. M.:* Nauchn. Dokl. Vysshei Shkoly Khim. i. Khim. Tekhnol. *1959*, 224; Chem. Abstr. *53*, 19493 (1959);
    b) *Vilkov, L. V., Rambini, N. G., Spiridonov, V. P.:* J. Struct. Chem. (USSR) (Engl. Transl.) *8*, 715 (1967).
83. *Pappalardo, R.:* J. Mol. Spetry. *29*, 13 (1969).
84. *Pappalardo, R.:* J. Chem. Phys. *49*, 1545 (1968).
85. *Bell, J. T., Thompson, C. C., Helton, D. M.:* J. Phys. Chem. *73*, 3338 (1969).
86. *Jørgensen, C. K.:* Modern aspects of ligand field theory. Amsterdam. North-Holland Publishing Company 1971.
87. *Löwdin, P-O.:* Advan. Phys. *5*, 1 (1956).
88. *Hart, F. A., Newberry, J. E.:* J. Inorg. Nucl. Chem. *31*, 1725 (1969).
89. *Boudreaux, E. A., Mukherji, A. K.:* Inorg. Chem. *5*, 1280 (1966).
90. *Mason, S. F., Peacock, R. D., Stewart, B.:* Chem. Phys. 'Letters *29*, 149 (1975): a full paper is in preparation.
91. *Stewart, B.:* Thesis, University of London (1975).
92. *Höhn, E. G., Weigang, O. E., Jr.:* J. Chem. Phys. *48*, 1127 (1968).
93. *Carlson, B. C., Rushbrooke, G. S.:* Proc. Cambridge Phil. Soc. *46*, 626 (1950).
94. *Siddall, T. H., Stewart, W. E.:* J. Inorg. Nucl. Chem. *32*, 1147, (1970).
95. *Weber, M. J.:* J. Chem. Phys. *48*, 4774 (1968).
96. *Blasse, G., Bril, A.:* Philips Res. Reports *21*, 379 (1966).
97. *Kajiura, M., Shinagawa, K.:* J. Phys. Soc. Japan *28*, 1041 (1970).
98. *Wybourne, B. G.:* In: Optical properties of ions in crystals (ed. *Crosswhite, H. M.* and *Moos, H. W.,*) p. 35. New York: Interscience 1967.
99. *Nieuwport, W. C., Blasse, G., Bril, A.:* In: Optical Properties of ions in crystals (ed. *Crosswhite, H. M.* and *Moos, H. W.,*) p. 161. New York: Interscience 1967.
100. *Choppin, G. R., Henrie, D. E., Buijs, K.:* Inorg. Chem. *5*, 1743 (1966).
101. *Birnbaum, E. R., Gomez, J. E., Darnall, D. W.:* J. Am. Chem. Soc. *92*, 5287 (1970).
102. *Karraker, D. G.:* Inorg. Chem. *6*, 1863 (1967).
103. *Karraker, D. G.:* Inorg. Chem. *7*, 473 (1968).
104. *Karraker, D. G.:* Inorg. Nucl. Chem. *31*, 2851 (1969).
105. *Karraker, D. G.:* J. Inorg. Nucl. Chem. *33*, 3713 (1971).
106. *Liu, C. S., Zollweg, R. J.:* J. Chem. Phys. *60*, 2384 (1974).
107. *Peacock, R. D., Weakley, T. J. R.:* J. Chem. Soc. A *1971*, 1937.
108. *Spedding, F. H., Rord, J. A.:* J. Phys. Chem. *78*, 1435 (1974).

109. *Seal, R. H.:* Thesis, University of London (1974).
110. *Gale, R., Godfrey, R. E., Mason, S. F.: Peacock, R. D., Stewart, B.:* J. C. S. Chem. Comm., in the press.
111. *Schwartz, R. W.:* to be published, personal communication to the author.
112. *Hearson, J. A.:* Thesis, University of London (1974).
113. *Peacock, R. D.:* Thesis, University of Dundee (1970).

Received October 29, 1974

Radiative and Non-Radiative Transitions of Rare-Earth Ions in Glasses

Renata Reisfeld

Department of Inorganic and Analytical Chemistry
The Hebrew University of Jerusalem, Israel

Table of Contents

I. Introduction

Since the publication of the Review on Spectra and Energy Transfer of Rare Earths in Inorganic Glasses (1), a great amount of experimental data has been accumulated on radiative and nonradiative relaxations of ions in glasses as well as on energy transfer from ions in which the absorption is allowed, to the rare earth ions.

At the same time, an increased activity has been observed in the field of similar phenomena of Rare Earth Ions in crystals and described in three excellent reviews published recently (2, 3, 4).

As a result of these works, we now have a much better understanding of the important questions such as ion lattice interaction and cooperative effects which are essential to elucidation of luminescence quantum yields.

Recently, there has been also a tremendous growth in the utilization of rare earth activated materials. These include not only phosphors, but also quantum electronic devices such as quantum counters, infrared to visible upconverters and lasers, the latter being of special interest in connection with nuclear fusion.

Because of these implications to glasses, it appears of interest to correlate experimental data with the theoretical approach of radiative transitions. To this end, a most useful approach is condensed in the Judd–Ofelt theory.

An important feature of the Judd–Ofelt theory (5, 6) is that once a set of three intensity parameters for a specific rare earth ion in a given host has been obtained, they can be used to calculate absorption and emission probabilities between any f^N levels of the system. This includes transitions such as excited state absorption which are difficult to measure experimentally.

Three main processes are active in the relaxation of rare earth ions from an excited level which may be populated either by radiative absorption or via energy transfer. These are:

1. Radiative decay
2. Nonradiative decay where the excitation energy is converted into vibrational quanta of the surroundings
3. Nonradiative transfer of energy between ions with possible degradation of excitation.

We shall summarize first the main features of the first two theories of these phenomena, then discuss our experimental data in view of these theories, so that deduction can be performed for new systems. The third part will be discussed in the near future.

Radiative Transition Probabilities: Optical line spectra of triply ionized rare earth ions originate from transitions between levels of the $4f^N$. The positions of these levels arise from a combination of the Coulomb interaction among the electrons, the spin orbit coupling and the crystalline electric field. The electrostatic interaction yields terms ^{2S+1}L with separations of the order 10^4 cm^{-1}. The spin orbit interaction then splits these terms into J states with typical splittings of 10^3 cm^{-1}. Finally, the J degeneracy of the free ion states is partially or fully removed by the crystalline Stark field, the width of a Stark manifold usually extending over several hundred cm^{-1}. In glasses, the amount of splitting is of the order of magnitude of inhomogeneous broadening as a result of the multiplicity of sites.

The free ion states obtained by diagonalizing the combined electrostatic and spin-orbit energy matrices are linear combinations of Russell-Saunders states of the form,

$$| f^N [\gamma SL] J > = \sum_{\gamma SL} c(\gamma SL) | f^N \gamma SLJ >$$

In this intermediate coupling scheme the total angular momentum J is a good quantum number, but the spin and orbital momentum numbers S and L are not (this is denoted by the brackets); γ includes whatever other quantum numbers are required to specify the states.

The rate of relaxation of an excited J state is governed by the combination of probabilities for radiative (A) and nonradiative (W) processes. The lifetime τ_a of an excited state a is given by,

$$\frac{1}{\tau_a} = \sum_b A_{ab} + \sum_b W_{ab} \tag{1}$$

where the summations are for transitions terminating on all final states b.

The radiative probability A includes both purely electronic and phonon assisted transitions. The nonradiative probability W includes relaxation by multiphonon emisison, and effective energy transfer rates arising from ion-ion interactions. The radiative quantum efficiency η_a is defined by,

$$\eta_a = \frac{\sum_b A_{ab}}{\sum_b A_{ab} + \sum_b W_{ab}} = \tau_a \sum A_{ab} \tag{2}$$

From the knowledge of any two of the quantities τ_a, η, $\sum A$ and $\sum W_{ab}$, the other two quantities may be determined.

Meaningful *ab initio* calculations of either electric dipole radiative or nonradiative decay rates are still beyond our present capabilities. Therefore they all involve experiment and phenomenological treatment.

In order to measure quantum yields, reliable standards of fluorescence are required. These have been measured in our laboratories for Rare Earth Ions (RE) and are summarized in Ref. (7).

II. Magnetic Dipole and Electric Quadrupole Transitions

The symmetry properties of the crystal field were invoked in early spectroscopic studies to account for the selection rules in the radiative transitions. With the availability of good energy level assignments, calculations of oscillator strengths for magnetic transitions and electric quadrupole transitions could be made.

Magnetic-dipole transitions (MD) in the trivalent rare earths are parity allowed between states of f^N. The selection rules for such transitions are

$$\Delta l = \Delta S = \Delta L = 0 \quad \text{and} \quad |\Delta J| \leqslant 1 \ (0 \not\leftrightarrow 0)$$

in the Russell-Saunders limit.

The line strength for MD transitions is given by,

$$S_{MD}(aJ;bJ') = \beta^2 \,|< f^N \,[\gamma SL]\, J\,|\,|\vec{L} + 2\vec{S}|\,|f^N \,[\gamma'\, S'\, L']\, J' >|^2 \qquad (3)$$

where $\beta = e\hbar/2mc$.

The matrix elements of magnetic-dipole operator $\vec{L} + 2\vec{S}$ between SLJ states may be calculated by formulae given by *Wybourne* (8) and have been tabulated for transitions from the ground state by *Carnall et al.* (9).

The spontaneous emission probability for magnetic dipole transitions is calculated using Eq. (3) and the usual formula for spontaneous emission probability given by,

$$A(aJ\ ;bJ') = \frac{64 \, \pi \, \nu^3 \, \chi_{MD}}{3(2\,J + 1)\, hc^3} \ S_{MD}(aJ;bJ') \qquad (4)$$

where χ is the local field correction for magnetic transitions,

$$\chi_{MD} = n^3 \ (n \text{ being the refractive index}).$$

Electric quadrupole transitions are also parity allowed between states f^N. The selection rules for such transitions are $\Delta S = 0$; $|\Delta L|$, $|\Delta J| \leqslant 2$. Possible enhancement mechanisms for quadrupole transitions have been considered by *Jørgensen* and *Judd* (10).

In glasses, the contribution of magnetic dipole and electric quadrupole to the transitions between various f^N multiplets compared to the forced electric dipole transitions is relatively weak because the rare earth ion is situated at a low symmetry site in a vitreous medium (7).

III. Forced Electric Dipole Transitions

The selection rules for electric dipole transitions are $\Delta l = \pm 1$ $\Delta S = 0$ $|\Delta L|$ and $|\Delta J| \leqslant 2l$ for RE $l = 3$.

Transitions between f^N levels involve no change in parity and are forbidden by the Laporte rule. They become allowed if odd harmonics in the static or dynamic crystal field admix states of opposite parity into the $4f$ level. In the *Judd* (5) *Ofelt* (6) theory the electric dipole matrix element for the ϱth component of polarization can be expressed as

$$< \psi_a \,|P_\varrho|\, \psi_b >= \sum_{q,t,\,\text{even}} Y(t, q, \varrho) < f^N \, \gamma SL \, JJ_z \,|U_{\varrho+q}^{(\lambda)}\,|\, f^N \gamma'\, S'\, L'\, J'\, J_z' > \qquad (5)$$

where the energy denominator $(E_{n'l'} - E_{4f})_{av}$, the crystal field parameters A_q^k odd and the interconfigurational radial integrals $\int R(4f) R(n'l') r^k d\gamma$ ($n'l'$ is the index for the next excited configuration) are all incorporated into the phenomenological parameters $Y(t, q, \varrho)$. The number and type of the component A_q^k which enter the $Y(t, q, \varrho)$ can be determined from group theory. [See Ref. (7)].

Application of this approach to the intensities of transitions between various Stark levels is dependent upon the availability of reliable crystal field eigenstates. When ions reside at low symmetry the states cannot be calculated and the intensity parameters have to be obtained from experimental data.

In an intermediate coupling scheme the line strength S for ED transitions reduces to a simple expression containing three intensity parameters given by,

$$S_{ED}(aJ; bJ') = S_{ED}(aJ: bJ')$$
$$= e^2 \sum_{\lambda=2,4,6} \Omega_\lambda |< f^N [\gamma SL] J||U^{(\lambda)}|| f^N [\gamma' S' L'] J' >|^2 \tag{6}$$

The free ion states obtained by diagonalizing the combined electrostatic and spin orbit energy matrices are linear combinations of the Russell-Saunders states of the form

$$| f^N [\gamma SL] J > = \sum_{\gamma SL} c(\gamma SL) | f^N \gamma SLJ >$$

and the matrix elements of $U^{(\lambda)}$ can be derived as described by *Wybourne* (8) using tabulated doubly reduced matrix elements of *Nielson* and *Koster* and 3 j and 6 j symbols (11). The oscillator strength p and the absorption cross section σ for a transition of frequency ν are related to S by

$$p(aJ; bJ') = [8 \pi^2 m\nu/3(2J + 1) he^2] S(aJ; bJ') \tag{7}$$

and

$$\int \sigma(\nu) d\nu = \frac{\pi e^2(n^2 + 2)^2}{9mcn} p \tag{8}$$

The oscillator strength of a transition between state $<f^N \psi J|$ and state $|f^N \psi' J'>$ in an isotropic medium is given by (9).

$$p = \frac{\sigma}{2J + 1} \sum_{\lambda=2,4,6} \tau_\lambda |< f^N \psi J||U^\lambda|| f^N \psi' J' >|^2 \tag{9}$$

A common form of (9) is with $T_\lambda = \dfrac{\tau_\lambda}{2J + 1}$ as follows:

$$p = \sigma \sum_{\lambda=2,4,6} T_\lambda |< f^N \psi J||U^\lambda|| f^N \psi' J' >|^2 \tag{10}$$

where σ (cm^{-1}) is the baricenter of the absorption band.

Experimentally, p is obtained from the absorption spectrum by

$$p = 4.318 \times 10^{-9} \int \varepsilon(\sigma) \, d\sigma \tag{11}$$

The quantity T_λ contains the sum of odd terms due to the static crystal field and the vibronic perturbation, the energy denominators which are the difference between the $4f$ and the next excited configuration, and the interconfigurational radial integrals $\int R(4f) \, R \, (n' \, l') \, r^k \, dr$.

For calculation of spontaneous emission probability the connection between A_{ed} and the measured calculated line strength is important.

$$A(aJ; bJ') \frac{64 \, \pi^4 \, \nu^3 \, \chi}{3(2J+1) \, hc^3} \, S(aJ: bJ') \tag{12}$$

This may be obtained from the relation:

$$A_{ed} = \frac{8\pi^2 e^2 \nu^2 \chi}{mc^3} \, p = \frac{8\pi^2 e^2 \sigma^2 \chi}{mc} \, p$$
$$= \frac{8\pi^2 e^2 \sigma^3 n^2}{mc(2J+1)} \sum_{\lambda=2,4,6} \tau_\lambda < f^N \psi \, J \, ||U^{(\lambda)}|| f^N \psi' \, J'> |^2 \tag{13}$$

where ν is the frequency of the transition in $\sec^{-1}$, χ is the field correction for electric dipole transition $\chi = \dfrac{n(n^2+2)^2}{9}$ and n is the refractive index.

The standard method of calculating the T_λ parameters (12) is to calculate the eigenvectors of the Rare Earth ion in a condensed system by diagonalizing the complete energy matrices using parameters which minimize the deviation between the centers of gravity of the observed and "free ion" electronic energy levels. The U^λ reduced matrix elements are taken from tables of *Nielson* and *Koster* (11) and transformed from L—S basic states to the physical (intermediate) coupling scheme, then squared and substituted into Eq. (10) for the oscillator strength.

As previously mentioned, the T_λ so obtained are the values which minimize the RMS (root mean square) deviation between the observed oscillator strengths and those calculated by means of Eq. (10).

Matrix elements of the tensor operator $||U^\lambda||$ for intermediate coupling have been tabulated for many ions and transitions. References to these calculations may be found in Ref. (2) (Table V—1). Although somewhat different sets of intermediate coupled eigenstates may have been used, the resulting values exhibit only small differences and hence values obtained for one host may generally be used for other hosts. This point has been discussed by *Reisfeld et al.* (13) and *Riseberg* and *Weber* (2).

It should be noted that the intensity parameters Ω_λ of Eq. (6) and the τ_λ parameters of Eq. (10) are related by,

$$\Omega_\lambda = \frac{3h}{8\pi^2 mc} \frac{n^2}{\chi} \tau_\lambda = 9.0 \times 10^{-12} \frac{n^2}{\chi} \tau_\lambda$$

Since they are used interchangably by various authors, attention must be drawn to the different forms.

IV. Intensity Parameters of Trivalent Rare Earth Ions in Glasses

As stated in the Introduction, knowledge of the intensity parameters and the availability of the matrix elements of intraconfigurational transitions of a given R. E. ion in its host are required for the calculation of specific radiative transition probabilities.

In view of this, intensity parameters were obtained from the experimentally measured oscillator strengths for a variety of rare earth ions in various oxide glasses in this laboratory and recently also by *Weber* and *Krupke* for Nd^{3+} in various oxide glasses.

In order to test the amount of sensitivity of the calculated matrix elements to a given host, matrix elements for various R. E. ions in several hosts were examined. These included Pr^{3+} (9, 12, 14), Tm^{3+} (9, 12, 15), Eu^{3+} (16, 17), Sm^{3+} (9), and Er^{3+} (5, 6, 18, 19).

The ion for which the greatest amount of data on the reduced matrix elements are available is Er^{3+}. Matrix elements for Er^{3+} in various hosts are presented in Table 1.

This table reveals that the average difference between the reduced matrix elements of Er^{3+} in various hosts is about $\pm 15\%$, which does not exceed the maximum experimental error for the values used to determine T_λ.

a) Intensity Parameters of Europium

The most extensive study of intensity parameters in glasses was performed on Eu^{3+}. The dependence of T_λ of Eu^{3+} as a function of the glass network forming ions is described in Ref. (13), and as a function of glass modifying ions in Ref. (20).

For this calculation, reduced matrix elements were used from the works of *Carnall et al.* (9, 12) for aqueous solutions. The dependence of the parameters on glass network former and glass network modifier is presented in Table 2. From the results it can be seen that the intensity parameters are intimately bound with the nature of the ligand field surrounding the R. E. ion.

In the Judd–Ofelt theory, the parity forbidden f–f transitions become allowed as electric dipole by admixing of configurations of opposite parity. The admixture may not necessarily be due to the pure $5d$ or $5g$ level of the R. E.

Recently, evidence has been given for the existence of covalent bonding of RE ion embedded in glass matrices with its surrounding oxygens (1, 21, 22).

When such bonding exists the molecular orbitals composed of RE and ligand wave functions will have no definite parity when the ion is positioned at a site lacking a center of symmetry. In such a case, the forbidden f–f transitions become possible. The amount of covalency can be inferred from the position of the charge transfer band. Quantitatively, this fact can be expressed by the nephelauxetic parameter β and the optical electronegativity χ_{opt} of a given ligand (glass). The nephelauxetic parameter of a glass (1) is defined according to *Jørgensen* (23) as

$$\beta = \frac{\sigma_f - \sigma}{\sigma_f} \qquad (14)$$

Table 1. Values of $\|U^\lambda\|$ for Er^{3+} in various matrices

$S'L'J'$	$\|U^{(2)}\|^2$				$\|U^{(4)}\|^2$				$\|U^{(6)}\|^2$			
	$ErCl_3$ (aq) (5,6)	$HClO_4$ (aq) (12)	Y_2O_3 (18)	LaF_3 (19)	$ErCl_3$ (aq) (5,6)	$HClO_4$ (aq) (12)	Y_2O_3 (18)	LaF_3 (19)	$ErCl_3$ (aq) (5,6)	$HClO_4$ (aq) (12)	Y_2O_3 (18)	LaF_3 (19)
$^4I_{13/2}$	—	0.0195	0.01948	0.0188	—	0.1173	0.11722	0.1176	—	1.4299	1.42776	1.4617
$^4I_{11/2}$	0.034	0.0291	0.03119	0.0259	0.00084	0.0004	0.00055	0.0001	0.3917	0.3969	0.39692	0.3994
$^4I_{9/2}$	0	0	0	0	0.2086	0.1856	0.19361	0.1452	0.0169	0.0122	0.01246	0.0064
$^4F_{9/2}$	0	0	0	0	0.5051	0.5275	0.51191	0.5655	0.4512	0.4612	0.44448	0.4651
$^4S_{3/2}$	0	0	0	0	0	0	0	0	0.2138	0.2230	0.21259	0.2285
$^2H_{11/2}$	0.7150	0.7326	0.6603	0.7056	0.41216	0.4222	0.38365	0.4109	0.0978	0.0927	0.08852	0.0870
$^4F_{7/2}$	0	0	0	0	0.1456	0.1467	0.14669	0.1467	0.6255	0.6280	0.61753	0.6273
$^4F_{5/2}$	0	0	0	0	0	0	0	0	0.2204	0.2237	0.22510	0.2237
$^4F_{3/2}$	0	0	0	0	0	0	0	0	0.1338	0.1256	0.13598	0.1204
$^4G_{11/2}$	—	0.8970	0.96705	0.9178	—	0.5123	0.54764	0.5271	—	0.1172	0.1232	0.1197
$^2K_{15/2}$	—	0.229	0.02394	—	—	0.0043	0.0045	—	—	0.0794	0.08274	—
$^2G_{9/2}$	0	0	0	—	0.0172	0.0157	0.23396	—	0.2312	0.2278	0.13115	—
$^2G_{7/2}$	—	0	0	—	—	0.0186	0.01416	—	—	0.1153	0.12466	—
$^2P_{3/2}$	—	—	0	0	—	0	0	0	—	—	0.01668	0.026
$^4G_{9/2}$	—	0	0	0	—	0.2436	0.05337	0.0511	—	0.1199	0.00023	0.0002
$^4D_{5/2}$	—	0	0	0	—	0	0	0	—	0.0267	0.02937	0.020
$^4D_{7/2}$	—	0	0	—	—	0.8917	0.8886	—	—	0.0291	0.02908	—

Table 2. T_λ Parameters of Eu^{3+} in various phosphate glasses

Change of glass network former	Change of glass network modifier														
	$T_2 \times 10^9$					$T_4 \times 10^9$					$T_6 \times 10^9$				
	Na or K	Mg	Ca	Sr	Ba	Na or K	Mg	Ca	Sr	Ba	Na or K	Mg	Ca	Sr	Ba
Phosphate	5.89	13.80	22.03	7.92	14.40	6.71	12.57	24.26	24.60	16.46	2.61	5.46	13.34	10.53	9.40
Silicate	10.9					5.90					1.47				
Germanate	22.1					2.88					6.96				

where σ is the wavenumber of maximum absorption of the ion in a given medium and σ_f the maximum absorption wavenumber of the free ion.

The optical electronegativity χ_{opt} can be calculated from the maxima of charge transfer band (24) by the equation,

$$\sigma = 30 \, kK \, [\chi_{opt}(\chi) - \chi_{uncor}(M)]$$

where χ_{uncor} is the electronegativity of the probe cation (for example Eu^{3+}).

The energies of the $5d$ and charge transfer states of RE ions in a glass were shown to be dependent on the glass network former in which this ion is situated (7). This energy dependence is expressed quantitatively by the values of χ_{opt} and β as shown above.

As seen from Table 2, the increase in covalency is accompanied by increase of T_λ parameters of Eu^{3+}. However, it should be noted that the covalency is not the sole factor influencing the T_λ parameters. These parameters increase also when the monovalent cation Na^+ is replaced by divalent cations in the phosphate glass (Table 2). As inferred from the charge transfer spectra of Eu^{3+}, $4f$—$5d$ spectra of Tb^{3+} and $^1S_0$—$^3P_1$ spectra of Pb^{++} (Table 3) which are insensitive to the change of the glass modifying cation (contrary to the glass-forming cation), there is no change of the electronegativity of the glass and of the covalency of the Eu^{3+} ion with the glass when changing the modifying cation. This is presented in Fig. 1.

From this we conclude that the nephelauxetic effect which is caused by the lowering of the excited state of the ion by the surrounding crystal field, is influenced mainly by the symmetric part of the crystal field. The change in the T_λ parameters in the absence of the nephelauxetic effect shows that the non-symmetric part of the field has a major influence on the parameters. The different modifier ions are responsible for the distortion of the oxygens around the RE ions lowering the site symmetry and increasing the probability for f—f transitions.

This physical picture can be interpreted by the theory of Judd-Ofelt in the following way:

The explicit expression for the T_λ is given by:

$$T_\lambda = \chi \, [8 \, \pi^2 \, m/3 \, h] \, (2\lambda + 1) \sum_{t,p} |A_{t,p}| \, {}^2 \Xi^2(t, \lambda) \tag{15}$$

In Eq. (15) the $A_{t,p}$ (t, odd) are the odd parity terms in the crystal field expansion depending on the site symmetry of the ion in a given matrix. The quantity $\Xi(t, \lambda)$ contains integrals involving the radial parts of the $4f^N$ wave functions and the excited opposite-parity electronic state wave functions. The energy difference between these two states appears in the denominator of $\Xi(t, \lambda)$.

$\Xi(t, \lambda)$ is a measure of the symmetric part of the crystal field (amount of covalency) and is the same for all the glasses with the same network modifier, depending only on the network former.

τ_λ parameter varies by the same amount with change of network former, as with change of network modifier. Hence, the $\Xi(t, \lambda)$ quantity which is not dependent on the network modifier is not the factor causing the change in the τ_λ's. As may be seen from Eq. (15), the remaining quantity responsible for the magnitude

Table 3. The ultraviolet excitation bands of Eu^{3+}, Tb^{3+} and Pb^{2+} in various matrices (in $kK = 1000$ cm^{-1})

Change of glass network former	Change of glass network modifier														
	Eu^{3+} (charge transfer)					$Tb^{3+}(4f \rightarrow 5d)$					$Pb^{2+}(^{1}S_0 \rightarrow {}^{3}P_1)$				
	Na or K	Mg	Ca	Sr	Ba	Na or K	Mg	Ca	Sr	Ba	Na or K	Mg	Ca	Sr	Ba
Phosphate	43.71	43.14	43.28	43.03	43.00	46.00	45.57	45.50	45.59	45.48	41.60	40.31	40.31	40.30	40.29
Borax	41.66					44.50					38.10				
Germanate	38.46					35.70					33.90				

133

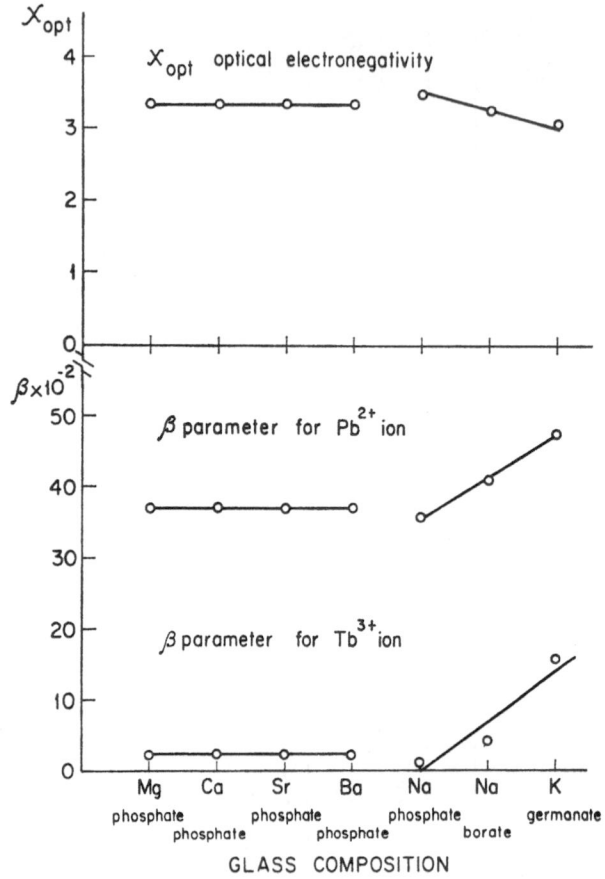

Fig. 1. Nephelauxetic parameter and optical electronegativity of various glasses as a function of glass composition

of the τ_λ parameter are the asymmetric terms in the crystal field expansion, the A_{tp} (t, odd). These remaining quantities arise, therefore, from the distortion of the cube surrounding the RE ions.

b) Samarium

The intensity parameters of Sm^{3+} in glasses were obtained from the experimentally measured oscillator strengths and matrix elements of *Carnall et al.* (12). It was shown by *Reisfeld et al.* (13) that the oscillator strengths of Sm^{3+} may be arranged in two groups, one referring to transitions up to 10,700 cm^{-1} and the second to transitions in the energy range 17,600—32,800 cm^{-1}. τ_λ parameters were calculated separately for these two regions. It was found that the parameters obtained for the high energy region may be higher by a factor of 30 in the parameter τ_2. For example, the values obtained for the Sm^{3+} ion in borate glass are as follows:

energy range in cm^{-1}	τ_2	τ_4	τ_6
low energy: 4520—10700	3.14×10^{-9}	6.00×10^{-9}	0.54×10^{-9}
high energy: 17600—32700	99.6×10^{-9}	7.2×10^{-9}	4.79×10^{-9}

This is not surprising since Eqs. (9) and (15) apply to the case where the f^N splittings are small compared to the f—d energy gap and therefore, it is incorrect to use oscillator strengths of transitions which are about 10,000 cm^{-1} for calculations of τ_λ parameters by means of the Judd-Ofelt Theory.

The τ_λ parameters of Sm^{3+} in various glasses are presented in Table 4.

Table 4. τ_λ Parameters for Sm^{3+} in various matrices

Matrix	Energy range (cm^{-1})	Number of data points	$\tau_2(cm)$	$\tau_4(cm)$	$\tau_6(cm)$
Phosphate low	4525—10,600	7	6.06×10^{-9}	4.17×10^{-9}	8.04×10^{-9}
Borate low	4520—10,700	8	3.14×10^{-9}	6.0×10^{-9}	0.54×10^{-9}
Germanate low	6000—1000	6	5.92×10^{-9}	4.96×10^{-9}	4.24×10^{-9}

The data of Sm^{3+} in germanate glasses are taken from the work of *Reisfeld, Bornstein* and *Boehm* (25).

Samarium was also studied in arsenic borax glasses (26). No absorption could be measured in the glasses because the host glass is opaque in the major part of the spectral range. The fluorescence of Sm^{3+} in the glasses increases with the amount of As_2O_3 while the fluorescent lifetime of the $^4G_{5/2}$ level decreases in the same order. This may be seen in Fig. 2. These phenomena may be explained by the increase of population of the excited states of Sm^{3+} in glasses containing As_2O_3 as a result of intermixing of these electronic levels of Sm^{3+} with the molecular orbitals of As_2O_3.

c) Erbium

Intensity parameters of Er^{3+} in borate, phosphate, germanate and tellurite glasses were obtained by *Reisfeld* and *Eckstein*, (27, 28). The results are presented in Table 5. The reduced matrix elements, $U^{(\lambda)}$ in these calculations between intermediate coupled states of Er^{3+} were evaluated by *Weber* (19).

In Ref. (29) radiative transition probabilities and quantum efficiencies of visible fluorescence of the $^4S_{3/2}$ and $^4F_{9/2}$ levels were measured by the comparative method and from measured decay times and the radiative transition probabilities of Eq. (12).

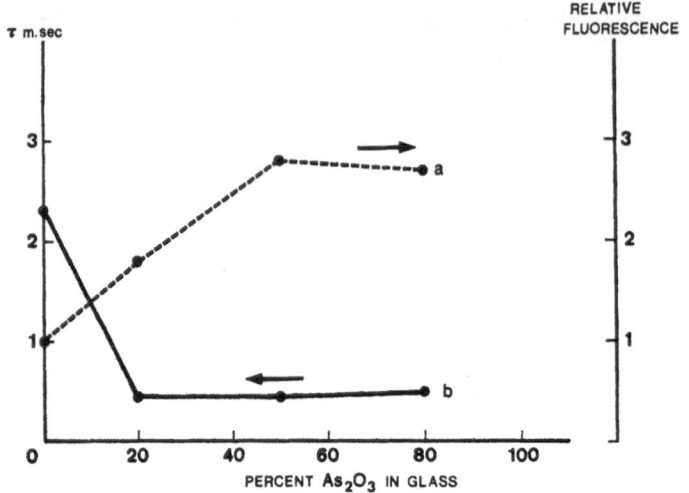

Fig. 2. (a) Relative fluorescence and (b) decay time of the $^4G_{5/2}$ level of Sm^{3+} as a function of As_2O_3 contained in arsenic borax glasses

Table 5. Values of τ_λ for Er^{3+} in various glasses

$\tau_\lambda \times 10^9$	Borate glass	Phosphate glass	Germanate glass	Tellurite glass
τ_2	16.25 ± 0.81	13.98 ± 0.66	13.31 ± 0.49	19.29 ± 0.37
τ_4	5.24 ± 1.30	5.27 ± 1.07	2.17 ± 0.78	3.38 ± 0.46
τ_6	3.20 ± 0.66	1.92 ± 0.54	0.97 ± 0.40	2.81 ± 0.20

In order to estimate the validity of such an approach, some of the results are presented below:

Table 6 presents the measured oscillator strengths of Er^{3+} in germanate and tellurite glasses, calculated by formula

$$P = 4.318 \times 10^{-9} \, \varepsilon(\sigma) \, d\sigma \tag{16}$$

For comparison, the oscillator strengths calculated by using Eq. (13) with the above τ_λ parameters and the matrix elements $U^{(\lambda)}$, are also included in Table 6. The degree of the fit is expressed by the root mean square deviation (RMS) between the observed and calculated P-values. The matrix elements for a given transition of Er^{3+} were evaluated by *Weber* for Er: LaF_3 (*19*).

The calculated radiative transition probabilities for various levels of Er^{3+} are tabulated in Table 7. Probabilities for electric dipole emission and total radiative lifetime, predicted for all transitions from $^4S_{3/2}$ and $^4F_{9/2}$ levels are also presented. The spontaneous emission probabilities, the total radiative lifetimes and

Table 6. Oscillator strength of Er^{3+}

$S'L'J'$	Spectral region	$P \times 10^6$ germanate		$P \times 10^6$ Tellurite	
	(cm^{-1})	Experimental	Calculated	Experimental	Calculated
$^4I_{13/2}$	6360— 6730	1.10	0.83	2.55	2.02
$^4I_{11/2}$	9900—10400	0.29	0.49	1.09	0.50
$^4I_{9/2}$	12400—12550	0.21	0.32	0.70	0.51
$^4F_{9/2}$	15100—15400	1.00	1.55	3.02	2.93
$^4S_{3/2}$	17800—18800	1.09	1.00	2.12	2.02
$^2H_{11/2}$	19100—19300	10.96	12.90	17.68	18.92
$^4F_{7/2}$	20000—20500	0.93	1.19	2.64	3.10
$\left.\begin{array}{l}^4F_{5/2}\\^4F_{3/2}\end{array}\right\}$	21800—22600	0.32	0.47	1.32	1.19
$^2H_{9/2}$	24450—24550	0.43	0.39	1.04	1.08
$^4G_{11/2}$	25700—26550	22.41	21.59	—	32.00
RMS deviation $\times 10^7$		2.33		1.36	

the measured lifetimes are also included in Table 7. Since the dipolar spontaneous emission probability is proportional to ν^3, only the higher frequency transitions were considered.

The fluorescent lifetimes of the $^4S_{3/2}$ state, measured for a tellurite glass containing 0.5 wt.% Er^{3+} and for a germanate glass containing 2 wt.% are included in Table 7, also. Both lifetimes decrease with increasing Er^{3+} content. For example, in tellurite glasses the decrease is from 34.7 μs for 0.5 wt.% Er^{+3} to 11.9 μs for 3.0 wt.% Er^{3+}. The dependence of lifetimes on concentration arises also from ion-ion interaction, according to the scheme,

$$^4G_{11/2} \longrightarrow {}^4F_{7/2} \rightsquigarrow {}^4I_{15/2} \longrightarrow {}^4I_{13/2}$$

$$^2H_{11/2},\ ^4S_{3/2} \longrightarrow {}^4I_{9/2} \rightsquigarrow {}^4I_{15/2} \longrightarrow {}^4I_{13/2}.$$

The apparent quantum efficiency (AQE) of the fluorescent transitions $^4S_{3/2} \rightarrow {}^4I_{15/2}$ at various excitation wavelengths were measured by a method described in detail in (7). In Table 8, the AQE's of the $^4S_{3/2} \rightarrow {}^4I_{15/2}$ and $^4F_{9/2} \rightarrow {}^4I_{15/2}$ transitions in tellurite and germanate glasses are presented. The probabilities of populating the fluorescent level via excitation to upper electronic levels, calculated by the following equation, [see Ref. (30)]:

$$Q(A) = P(A)\ Q(^4S_{3/2} \text{ or } ^4F_{9/2}) \tag{17}$$

Table 7. Reduced matrix elements, intensity parameters, τ_λ, and predicted probabilities for spontaneous emission from $^4S_{3/2}$ and $^4F_{9/2}$ levels of Er^{3+}, in germanate and tellurite glasses

Reduced matrix elements between intermediate states of Er^{3+}

[SL] J	[S′L′] J′	$[U^{(2)}]^2$	$[U^{(4)}]^2$	$[U^{(6)}]^2$	Germanate glass A_{ed} (s^{-1})	$(\sum A_{ed})^{-1}$ $=\tau_{nat}$ (μs)	τ_{eff} (μs)	τ_{meas} (μs)	Tellurite glass A_{ed} (s^{-1})	$(\sum A_{ed})^{-1}$ $=\tau_{nat}$ (μs)	τ_{eff} (μs)	τ_{meas} (μs)
$^4F_{9/2}$	$^4I_{15/2}$	0.0	0.5655	0.4651	676.4				3021.0			
	$^4I_{13/2}$	0.0096	0.1576	0.0870	206.8	1060.0	—	—	470.6	272.0	—	—
	$^4I_{11/2}$	0.0671	0.0088	1.2611	51.5				173.4			
	$^4I_{9/2}$	0.0960	0.0061	0.0120	8.6				11.6			
$^4S_{3/2}$	$^4I_{15/2}$	0.0	0.0	0.2285	2668.6				7849.0			
	$^4I_{13/2}$	0.0	0.0	0.3481	1073.9	253.0	159.0	8.9	2774.5	90.16	64.7	34.7
	$^4I_{11/2}$	0.0	0.0037	0.0789	93.3				206.4			
	$^4I_{9/2}$	0.0	0.0792	0.2560	142.1				261.2			

Table 8. Apparent quantum efficiencies (AQE) of Er $^4S_{3/2}$ and $^4F_{9/2}$ emissions and probability of populating the $^4S_{3/2}$ and $^4F_{9/2}$, (P(A)), levels via excitation to upper selected levels in germanate and tellurite glasses

Emissions from	Excited level	AQE germanate	AQE tellurite	P(A) germanate	P(A) tellurite
$^4S_{3/2}$	$^4S^3/_2$	0.0200	0.317	1.000	1.000
	$^2H_{11/2}$	0.0121	0.296	0.636	0.933
	$^4F_{7/2}$	0.0066	0.122	0.346	0.325
	$^4F_{5/2,3/2}$	0.0067	0.103	0.351	0.326
	$^2H^9/_2$	0.0097	0.200	0.523	0.631
	$^4G^{11}/_2$	0.0044	0.100	0.228	0.315
	$^2G^7/_2$	0.0097	—	0.247	—
$^4F_{9/2}$	$^4F_{9/2}$	0.0023	0.0127	1.000	1.000
	$^4S_{3/2}$	0.0021	0.0076	0.950	0.600
	$^2H_{11/2}$	0.0010	0.0076	0.420	0.600
	$^4F_{7/2}$	0.0009	0.0114	0.410	0.900
	$^4F_{5/2,3/2}$	—	0.0056	—	0.440
	$^2H_{9/2}$	—	0.0058	—	0.463
	$^4G_{11/2}$	0.0010	0.0028	0.130	0.220
	$^2G_{7/2}$	—	0.0032	—	0.255

are also included in Table 8. In Eq. (17) the Q(A) is the apparent fluorescent efficiency determined upon excitation of level A, Q($^4S_{3/2}$ or $^4F_{9/2}$) is the apparent fluorescent efficiency calculated when $^4S_{3/2}$ or $^4F_{9/2}$ are excited directly and P(**A**) is the probability that an excited ion initially in state A will be converted to the $^4S_{3/2}$ or $^4F_{9/2}$ levels.

Apart from the AQE's, the total quantum efficiencies (η) of Er $^4S_{3/2}$ and $^4F_{9/2}$ fluorescences were determined. For these determinations radiative transition probabilities from $^4S_{3/2}$ or $^4F_{9/2}$ to $^4I_{13/2}$, $^4I_{11/2}$, and $^4I_{9/2}$ are needed in addition to the measured intensities of $^4S_{3/2}$ and $^4F_{9/2} \rightarrow {}^4I_{15/2}$ transitions. The radiative probability of the measured $^4S_{3/2} \rightarrow {}^4I_{15/2}$ and $^4F_{9/2} \rightarrow {}^4I_{15/2}$ transitions can be calculated using the formula [Ref. (7)]:

$$A_{ed} = 2.880 \times 10^{-9} \frac{g_l}{g_u} n^2 \nu^2 \int \varepsilon(\nu) \, d\nu \,, \tag{18}$$

assuming that:

$$A({}^4I_{15/2} \longrightarrow {}^4S_{3/2}) = A({}^4S_{3/2} \longrightarrow {}^4I_{15/2})$$

and

$$A({}^4I_{15/2} \longrightarrow {}^4F_{9/2}) = A({}^4F_{9/2} \longrightarrow {}^4I_{15/2})$$

where g_l and g_u are the degeneracies of the lower and upper states respectively, n is the index of refraction (1.65 for germanate and 2.15 for tellurite glasses) and $\varepsilon(\nu)$ is the extinction coefficient as a function of wave number. Using the calculated values of transition probabilities A $(^4S_{3/2} \rightarrow {}^4I_{15/2})$ and $A(^4F_{9/2} \rightarrow {}^4I_{15/2})$, obtained from the absorption spectra, and the predicted radiative transition probabilities from $^4S_{3/2}$ or $^4F_{9/2}$ to all terminal levels (see Table 8), the total radiative probability was found. The total quantum yield of the emitting level was determined from the relation:

$$\eta_i = Q_i \sum A^r_{ij}/A^r_{ig} \tag{19}$$

where Q_i is the AQE of the radiative transition from level i to the ground level, g, and A_{ig} is the radiative transition rate from i to g.

The total quantum efficiency can also be calculated from the measured and predicted total radiative lifetimes. In all the calculations of $\eta(^4S_{3/2})$ the effect of thermalization of $^2H_{11/2}$ must be taken into account (31). Since the radiative transition rate of $^2H_{11/2}$ is significantly different than that of $^4S_{3/2}$, the effective decay time will be different from the predicted lifetime of $^4S_{3/2}$ and in the absence of any nonradiative transitions to other lower levels, it is given by:

$$\frac{1}{\tau_{eff}} = \sum_i g_i A_i \{\exp - (E_i - E_S)/kT\} / \sum_i g_i \{\exp - (E_i - E_S)/kT\}, \tag{20}$$

where A_i is the total spontaneous emission probability from level i to all lower levels and g_i is the level degeneracy. The total quantum efficiencies of $^4S_{3/2}$, as obtained by the two methods, for Er-doped tellurite and germanate glasses, are presented in Table 9.

d) Thulium

In Ref. (32) calculations of radiative transitions for the states $^3P_0$, $^1I_6$, $^1D_2$, and $^1G_4$ of Tm^{3+}, using the intermediate coupling scheme described by *Weber* (19), were performed. The matrix elements of the unit tensor operator $U^{(\lambda)}$ between $^3P_0$, $^1I_6$, $^1D_2$, $^1G_4$ and all the next lower lying states were calculated using the eigenstates of Tm^{3+}, calculated by *Krupke* and *Gruber* (15) in an inter-mediate coupling. A set of $U^{(\lambda)}$ values for $\lambda = 2, 4, 6$ are given in Table 10.

The spontaneous emission probabilities for electric-electric dipole transitions from the $^3P_0$, $^1I_6$, $^1D_2$, and $^1G_4$ levels of Tm^{3+} in different glasses were calculated using the above matrix elements and the intensity parameters, τ_λ, obtained from the experimentally measured absorption spectra by the relationship given in Eq. (13). The τ_λ values for Tm^{3+} in different glasses are tabulated in Table 11.

The probabilities for magnetic-dipole transitions were not considered since they are generally small. The spontaneous transition from Tm^{3+} $^3P_0$, $^1I_6$, $^1D_2$ and $^1G_4$ levels to all lower levels, the total spontaneous emission probabilities, the predicted radiative lifetimes and the measured lifetimes are summarized in Table 12.

Table 9. Quantum efficiency of Tm ($^3P_0$, $^1I_6$), $^1D_2$ and of Er $^4S_{3/2}$, $^4F_{9/2}$ in various glasses, obtained by a comparison to quantum yield standard (Method I) and from measurements of lifetimes (Method II)

RE^{3+}	Excited state	Method I								Method II							
		Borate		Phosphate		Germanate		Tellurite		Borate		Phosphate		Germanate		Tellurite	
		80 K	300 K	80 K	300 K	80 K	300 K	80 K	300 K	80 K	300 K	80 K	300 K	80 K	300 K	80 K	300 K
Tm^{3+}	$^3P_0$, $^1I_6$										0.46		0.6				
	$^1D_2$	0.53	0.50	0.96	0.8	0.89	0.75	0.91	0.09	0.51	0.47	0.90	0.8	0.94	0.92	0.96	
Er^{3+}	$^4S_{3/2}$					0.050	0.044	0.614	0.590					0.049	0.046	0.54	0.53
	$^4F_{9/2}$					0.005	0.003	0.023	0.015								

Table 10.[a]) Squared reduced matrix elements of Tm^{3+}

(SL) J	[S′ L′] J′	$[U^{(2)}]^2$	$[U^{(4)}]^2$	$[U^{(6)}]^2$
$^3P_0$	$^3H_6$	0.	0.	0.0761
	$^3H_4$	0.	0.2757	0.
	$^3H_5$	0.	0.	0.
	$^3F_4$	0.	0.0195	0.
	$^3F_3$	0.	0.	0.
	$^3F_2$	0.3601	0.	0.
	$^1G_4$	0.	0.0513	0.
	$^1D_2$	0.0278	0.	0.
	$^1I_6$	0.	0.	0.0230
$^1I_6$	$^3H_6$	0.01154	0.04229	0.01460
	$^3H_4$	0.0646	0.5224	0.4096
	$^3H_5$	0.0010	0.0022	0.0061
	$^3F_4$	0.0728	0.3376	0.1069
	$^3F_3$	0.	0.0030	0.0080
	$^3F_2$	0.	0.0411	0.3528
	$^1G_4$	0.2050	1.2013	0.6050
$^1D_2$	$^3H_6$	0.	0.3133	0.0934
	$^3H_4$	0.5680	0.0928	0.0230
	$^3H_5$	0.	0.0011	0.0193
	$^3F_4$	0.1248	0.0096	0.2280
	$^3F_3$	0.1633	0.0687	0.
	$^3F_2$	0.0643	0.3065	0.
	$^1G_4$	0.1874	0.1799	0.0022
$^1G_4$	$^3H_6$	0.0464	0.0747	0.0100
	$^3H_4$	0.0020	0.0182	0.0693
	$^3H_5$	0.0773	0.0078	0.5633
	$^3F_4$	0.1645	0.0052	0.4114
	$^3F_3$	0.0117	0.0808	0.3253
	$^3F_2$	0.0095	0.0784	0.0432

[a]) The author is very grateful to Mr. John A. Caird from the Center for Laser Studies, University of Southern California, Los Angeles, for providing the matrix elements which were also obtained by him.

Table 11. Values of τ_λ for Tm^{3+} in various glasses

$\tau_\lambda \times 10^9$	Borate glass	Phosphate glass	Germanate glass	Tellurite glass
τ_2	8.65 ± 1.92	8.81 ± 1.13	5.75 ± 1.64	9.04 ± 1.25
τ_4	4.44 ± 0.81	4.32 ± 0.35	2.24 ± 0.69	4.70 ± 0.53
τ_6	2.65 ± 0.59	1.06 ± 0.27	1.16 ± 0.50	2.58 ± 0.38

The oscillator strengths of Tm^{3+} and Er^{3+} in T, B and P are higher than in G glasses. Since most of the transitions are of electric-dipole nature (with the exception of Tm^{3+} $^3H_6 \rightarrow {}^3H_5$ and Er^{3+} $^4I_{15/2} \rightarrow {}^4I_{13/2}$, $^2K_{15/2}$), these results imply that the non-symmetric component of the electric field acting on the rare earth in the T, B and P glasses is stronger than in G glasses.

Fluorescence was observed from the $^3P_0$, $^1I_6$, $^1D_2$ and $^1G_4$ levels of thulium in borate, phosphate and germanate and from $^1D_2$ and $^1G_4$ levels only in Tm^{3+} doped tellurite glasses. The fluorescence from the $^2H_{11/2}$, $^4S_{3/2}$ levels of Er^{3+} was very intense in tellurite and much weaker in the germanate glasses.

The fluorescence of Tm^{3+} in tellurite glass can be observed also by direct excitation of the tellurite matrix in the range 350—420 nm, as can be seen from the excitation spectra of the $\sim$455 nm ($^1D_2 \rightarrow {}^3H_4$) and of the $\sim$655 nm ($^1G_4 \rightarrow {}^3H_4$) emissions of thulium presented in Fig. 3. The appearance of the broad band in the range 350—420 nm, in addition to the $\sim$360 nm ($^1D_2$) band of Tm^{3+}, in the excitation spectra implies existence of energy transfer from the tellurite matrix to the thulium ions. No such band was observed in the excitation spectrum of Er^{3+} doped tellurite glass.

Temperature dependence of visible fluorescence of Tm^{3+} doped B, P, G and T glasses (Borate, phosphate, germanate and tellurite).

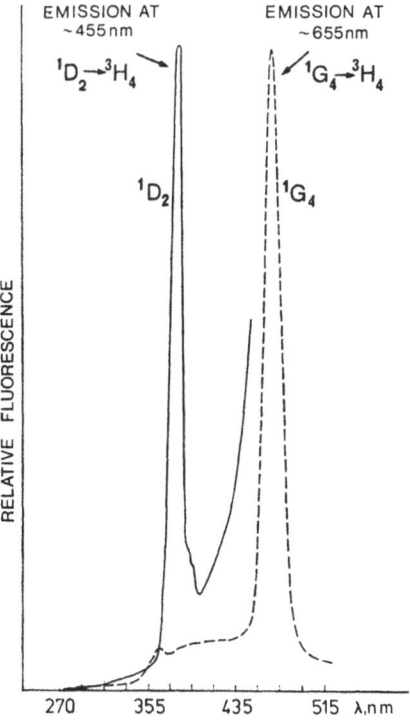

Fig. 3. Excitation spectra of $^1D_2 \rightarrow {}^3H_4$ ($\sim$455 nm) and $^1G_4 \rightarrow {}^3H_4$ ($\sim$655 nm) emissions of Tm^{3+} in tellurite glass

Table 12. Predicted probabilities for electric dipole spontaneous emissions from different levels of Tm^{3+} in glasses

		Borate				Phosphate				Germanate				Tellurite			
SLJ	S'L'J'	A_{ed}	$\sum A_{ed}$	τ_{rad} μsec	τ_{meas} μsec	A_{ed}	$\sum A_{ed}$	τ_{rad} μsec	τ_{meas} μsec	A_{ed}	$\sum A_{ed}$	τ_{rad} μsec	τ_{meas} μsec	A_{ed}	$\sum A_{ed}$	τ_{rad} μsec	τ_{meas} μsec
$^3P_0$	$^3H_6$	13447.9				5379.2											
	$^3H_4$	47641.5				46353.8											
	$^3F_4$	1481.2	101803	9.8		1441.1	92936.2	10.7	9.65								
	$^3F_2$	38118.1				38823.2			(300 K)								
	$^1G_4$	947.5				921.9											
	$^1D_2$	166.4				17.0											
$^1I_6$	$^3H_6$	1925.4			4.43	1769.2											
	$^3H_4$	14037.8			(300 K)	11545.9			(300 K)								
	$^3H_5$	97.9				70.1											
	$^3F_4$	3951.6	25567.2	39.1		3625.9	21254.8	47.0									
	$^3F_3$	44.9				27.9											
	$^3F_2$	1342.5				662.5											
	$^1G_4$	3930.6				3445.3											
	$^1D_2$	236.4				106.9											

$^1D_2$	31642.1	30186.6	22477.2	64766.2
	31.6	33.1	44.5	15.4
	14.7 (80 K)	20.0 (80 K)	26.4 (80 K)	9.6 (80 K)
$^3H_6$	10445.1	9258.6	6176.6	21408.4
$^3H_4$	16943.1	17079.2	13315.7	34713.5
$^3H_5$	126.9	57.1	62.8	244.1
$^3F_4$	1714.7	1382.5	1183.7	3431.5
$^3F_3$	1130.9	1142.7	870.7	2321.9
$^3F_2$	1097.5	1082.3	731.7	2268.7
$^1G_4$	183.9	184.3	136.0	378.2
$^1G_4$	2630.5	2087.0	1742.1	5277.7
	380.1	479.1	547	189.5
$^3H_6$	1234.6	1206.0	876.6	2536.0
$^3H_4$	196.9	110.7	113.3	387.3
$^3H_5$	858.5	512.7	527.6	1677.1
$^3F_4$	255.0	191.8	174.6	507.1
$^3F_3$	65.2	39.3	37.3	12.8
$^3F_2$	20.3	17.5	12.7	4.1

d) Thulium

The fluorescence of thulium doped tellurite glass at 80 K and 300 K, due to the $^1D_2 \rightarrow {}^3H_4$ (~455 nm) transition excited into the $^1D_2$ level and due to the $^1G_4 \rightarrow {}^3H_4$ (~655 nm) transition at $^1G_4$ excitation are presented in Fig. 4. The increase of the ~455 nm emission of thulium at low temperatures is by a factor of about 10, while the increase in the ~655 nm emission is by a factor of about 1.5 only, compared to the intensities at room temperature. These differences arise from the temperature dependence of the absorption edge of the tellurite host and the self absorption of the ~455 nm light by the matrix, while at ~655 nm the self absorption is negligible at all temperatures.

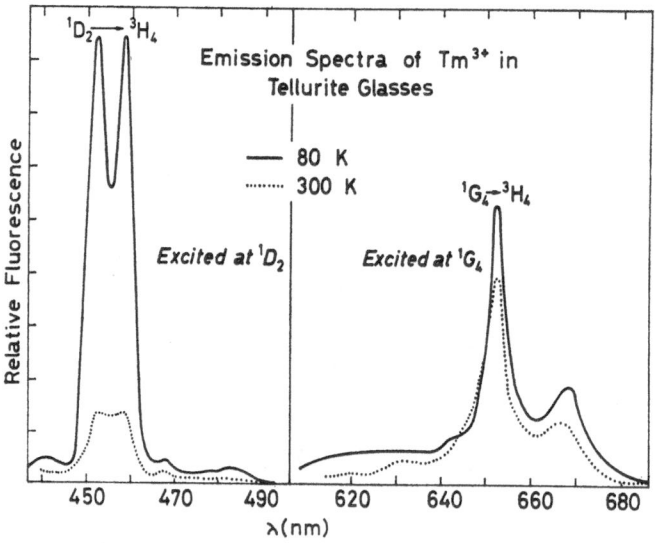

Fig. 4. Fluorescence spectra of Tm^{3+} in tellurite glass at 80 K and 300 K due to the $^1D_2 \rightarrow {}^3H_4$ transition excited into the $^1D_2$ level and due to the $^1G_4 \rightarrow {}^3H_4$ transition at $^1G_4$ excitation

The dependence of the $^1D_2 \rightarrow {}^3H_4$ emission intensity excited at $^1D_2$ (~360 nm) as a function of temperature for the B, P, G and T glasses is presented in Fig. 5 and that of the $^1G_4 \rightarrow {}^3H_4$ fluorescence is presented in Fig. 6. From Fig. 5 we see that there is only a mild decrease of the $^1D_2 \rightarrow {}^3H_4$ fluorescence in B, P and G glasses between 80 K and 280 K, but a very strong decrease in this fluorescence of Tm^{3+} in the tellurite. From the calculated multiphonon relaxation rates, which will be presented in the next chapter, we conclude that the strong decrease of the Tm^{3+} $^1D_2 \rightarrow {}^3H_4$ fluorescence with increase of temperature cannot be explained by multiphonon relaxation only, and must result from radiation trapping of the $^1D_2 \rightarrow {}^3H_4$. In order to test this assumption the absorption spectra of the four undoped glasses as a function of temperature were measured. The absorption spectra at room temperature are presented in Fig. 7. In general, the absorption edge of glasses is shifted to longer wavelengths as the temperature increases. This is especially significant in T glasses and to some extent in G glasses.

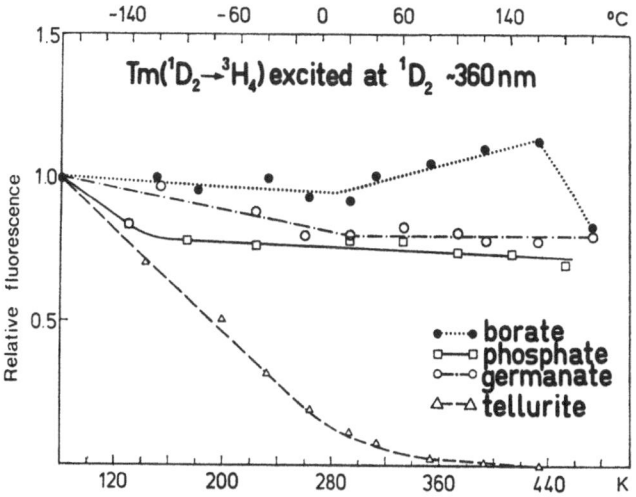

Fig. 5. Dependence of the $^1D_2 \rightarrow {}^3H_4$ ($\sim$455 nm) emission intensity of Tm^{3+} excited at $^1D_2$ ($\sim$360 nm) level on temperature for different glass media

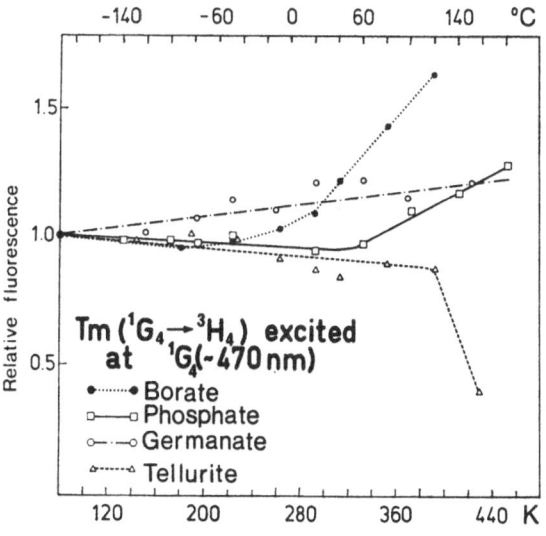

Fig. 6. Dependence of $^1G_4 \rightarrow {}^3H_4$ ($\sim$655 nm) emission of Tm^{3+} excited at $^1G_4$ ($\sim$470 nm) level on temperature in different glass media

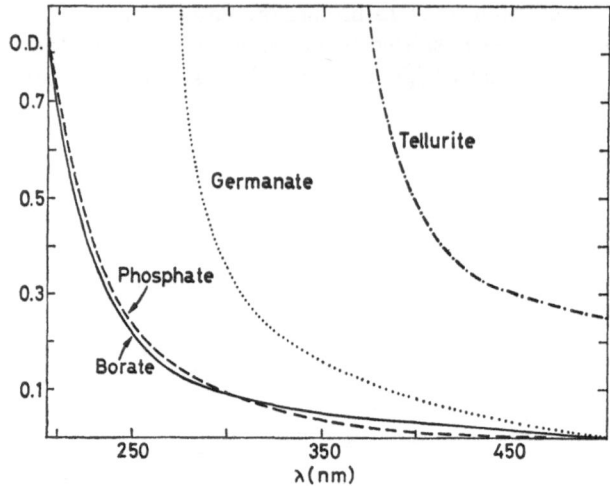

Fig. 7. Absorption edge of different glasses at room temperature

The difference in the temperature dependence of the fluorescence from the $^1D_2$ and the $^1G_4$ levels in T glasses, as shown in Figs. 5 and 6 respectively, may also be explained by the fact that (in spite of the similar energy gap between these and the next lower levels), the T matrix absorbs the $^1D_2 \rightarrow {}^3H_4$ ($\sim$455 nm) fluorescence, but not the $^1G_4 \rightarrow {}^3H_4$ ($\sim$655 nm) fluorescence. Hence, in the first case the temperature dependence of fluorescence is mainly due to the matrix absorption, while in the latter, we observe only a small multiphonon relaxation rate. The multiphonon relaxation from $^1D_2$ and $^1G_4$ levels should be of the same order of magnitude, since the energy gaps and the orbit-lattice coupling constant are similar in both cases.

The increase in fluorescence in the temperature range 280—440 K, as seen in Fig. 6 for B and P glasses and to a lesser extent for G glasses, may be explained by two alternative ways:

1. From the experimental fact that the transparency of the glass matrix increases in this range, permitting more rare earth centers to be excited (non-radiative transition would decrease the fluorescence with increasing temperature). This was also the explanation given by *Wanmaker* and *Bril* (*33*) for the enhanced brightness of Tb^{3+} $^5D_4 \rightarrow {}^7F_j$ fluorescence with increase of temperature. In Fig. 5, this effect is seen to a lesser extent for borate glasses only. Again the transparency of B glasses is higher in the 280—400 K temperature range at 360 nm, whereas the transparency of the P and G glasses in this range of wavelength and temperature is practically unaltered and that of T glasses even decreases;

2. By the energy transfer from borate glasses to Tm^{3+}. This assumption is drawn from the fact that the undoped borate glass exhibits an intrinsic broad fluorescence in the range of 600—700 nm with an excitation broad band peaking at 400 nm. The origin of this fluorescence is yet unknown (see Fig. 8). On addition of Tm^{3+} and with increase of temperature this fluorescence is diminished, implying an energy transfer from borate to Tm^{3+}.

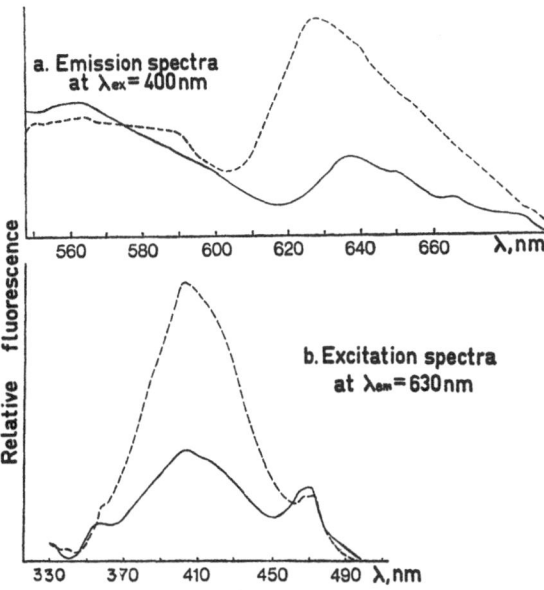

Fig. 8. Emission spectrum of undoped borate glass at 300 K (---) and 413 K (——)

The fluorescent quantum efficiencies of Tm^{3+} and Er^{3+} at 80 K and 300 K in different glass media were determined by the use of two alternative methods:

a) From the calculated transition probabilities and the measured decay times, using the relationship:

$$\eta_i = \frac{\sum A_{ij}}{\sum A_{ij} + \sum W_{ij}} = \tau_{meas} \sum A_{ij} \qquad (21)$$

where η_i is the quantum yield of the emitting level, i, $\sum A_{ij}$ — is the total probability for radiative decay from level i to all lower states $\sum W_{ij}$ denotes the probability for nonradiative decay and τ_{mes} — is the measured lifetime of level i. The A_{ij} — probabilities for electric dipole emissions — were calculated by use of τ_λ values (Table 5) and matrix elements $U^{(\lambda)}$ according to Eq. (13) The branching ratios β_{ij} for emission from state i to state j given by equation,

$$\beta_{ij} = A_{ij}/\sum_j A_{ij} \qquad (22)$$

for Tm^{3+} emissions from $^1D_2$ and for Er^{3+} emissions from $^4S_{3/2}$ and $^4F_{9/2}$ in different glass hosts are tabulated in Table 13.

b) The η can be calculated also by a comparison with a fluorescent standard as described in (1), using 0.1 wt.% Tb^{3+} doped borate glass for determination of ϕ_i of the Er^{3+} $^4S_{3/2}$ ($\sim$550 nm) and Tm^{3+} $^1D_2$ ($\sim$360 nm) levels while for the determination of the quantum efficiency of the Er^{3+} $^4F_{9/2}$ ($\sim$660 nm) level

Table 13. Branching ratio, β, of radiative transition from Tm $^1D_2$ and Er $^2H_{11/2}$, $^4S_{3/2}$ and $^4F_{9/2}$ levels in different glasses

Ion	Excited level	Terminal level	Branching ratio of transition in:			
			Borate	Phosphate	Germanate	Tellurite
Tm^{3+}	$^1S_2$	$^3H_6$	0.3301	0.3067	0.2748	0.3305
		$^3H_4$	0.5355	0.5658	0.5924	0.5360
		$^3H_5$	0.0040	0.0019	0.0028	0.0038
		$^3F_2$	0.0347	0.0359	0.0326	0.0350
		$^3F_3$	0.0357	0.0378	0.0387	0.0359
		$^3F_4$	0.0542	0.0458	0.0527	0.0530
		$^1G_4$	0.0058	0.0061	0.0060	0.0058
Er^{3+}	$^2H_{11/2}$	$^4I_{15/2}$			1.0000	1.0000
	$^4S_{3/2}$	$^4I_{15/2}$			0.6710	0.7080
		$^4I_{13/2}$			0.2700	0.2500
		$^4I_{11/2}$			0.0230	0.0190
		$^4I_{9/2}$			0.0360	0.0230
		$^4F_{9/2}$				
	$^4F_{9/2}$	$^4I_{15/2}$			0.7170	0.8220
		$^4I_{13/2}$			0.2190	0.1280
		$^4I_{11/2}$			0.0550	0.0470
		$^4I_{9/2}$			0.0090	0.0030

0.2 wt.% Eu^{3+} doped phosphate glass was used. The standards were chosen in such a way that the excitation and emission wavelengths of the standards and Tm^{3+} or Er^{3+} corresponded.

The quantum yields measured by method (a) and/or (b) are summarized in Table 9. As can be seen the quantum efficiencies are lowest in the borate glass, higher in phosphate and germanate and the highest in tellurite glasses.

Since, as seen from Tables 13, the radiative transition probabilities of Tm^{3+}, and (29), the radiative transition probabilities of Er^{3+} do not differ significantly in the different glass hosts, the differences in the observed quantum efficiencies should be due to the different non-radiative losses, namely different multiphonon relaxation rates.

e) Holmium

The radiative transition probabilities of Ho^{3+} in calibo glass, phosphate glass and tellurite glass have been obtained recently by *Reisfeld* and *Hormodaly* (34). The absorption spectra were measured in the visible and near IR part of the spectrum in tellurite glasses and in UV visible and IR part of the spectrum in phosphate and calibo glass. The oscillator strengths of Ho^{3+} in the glasses together with those of Ho^{3+} in YAlO$_3$ (35) are presented in Table 14. A set of Judd–Ofelt intensity

Table 14. Oscillator strength of Ho^{3+}

Assignments(35)	Wavenumber (cm^{-1})	Oscillator strength $\times 10^6$			
		Phosphate	Calibo	Tellurite	$YAlO_3$(35)
$^5I_8 \rightarrow {}^5I_7$	4717— 5555	—	1.544	1.952	1.28
$^5I_6$	8064— 9090	0.632	0.932	0.999	0.64
$^5I_5$	10638—11627	—	0.245	0.240	0.12
$^5I_4$	—	—	—	—	0.04
$^5F_5$	14880—16129	2.650	3.686	4.554	2.17
$^5S_2$ $^5F_4$	17857—19230	3.937	4.991	5.932	2.70
$^5F_3$	20000—21276	1.333	1.718	1.850	0.65
$^5F_2$ $^3K_8$	21052—21739	1.4182	1.341	0.998	1.09
$^5G_6$	21276—22988	20.470	26.504	41.270	7.45
$^5G_5$	23255—25000	2.828	3.591	—	2.10
$^5G_4$ $^2K_7$	25064—26666	0.514	1.039	—	0.51
$^3H_6$ $^3H_5$ $^5G_2$	26666—28238	5.964	7.4264	—	2.52
$^3L_9$ $^5G_3$	28238—29412	1.3091	1.502	—	0.49
$^3K_6$ $^3F_4$	29412—30769	1.391	1.375	—	0.54
$^3L_8$ $^3M_{10}$	33333—35088		4.810		0.18
$^5D_4$ $^3G_3$ $^3H_4$ $^3F_2$ $^1L_8$ $^3G_5$	35088—37037		3.050		4.15
$^3P_2$ $^3L_7$ $^3I_7$ $^3F_4$ $^3I_5$ $^3M_9$	37735—38910		0.798		0.15
$^3I_6$ $^3D_1$ $^5D_1$ $^3P_1$ $^5D_3$	39215—40485		0.538		
$^3F_4$ $^5D_4$	40485—41666		4.981		—

parameters was obtained from the observed oscillator strengths and Eq. (10), using a least-squares fitting procedure. Since magnetic dipole transitions make large contributions to the $^5I_8$–$^5I_7$ oscillator strengths, this value was deleted in fitting the electric dipole parameters. The $U^{(\lambda)}$ matrix elements for Ho^{3+} in LaF_3 were evaluated by *Weber* and *Matsinger* (*35*), using the "free ion" radial parameters of *Caspers et al.* (*36*). As mentioned earlier the matrix elements exhibit only small variations with host lattice, therefore, matrix elements of Ho^{3+} in LaF_3 were used for the glasses. The electric dipole intensity parameters Ω of Ho^{3+} in the glasses and $YAlO_3$ are presented in Table 15.

Table 15. Intensity parameters of Ho^{3+}

Host	$\Omega_2(10^{-20}\ cm^2)$	$\Omega_4(10^{-20}\ cm^2)$	$\Omega_6(10^{-20}\ cm^2)$
Phosphate glass	5.60	2.72	1.87
Calibo glass	6.83	3.15	2.53
Tellurite glass	6.92	2.81	1.42
$YAlO_3$(*35*)	1.82	2.38	1.53

Weak visible fluorescence of Ho^{3+} has been observed in phosphate glasses arising from the ($^5S_2$, $^5F_4$) and $^5F_5$ to the terminal $^5I_8$ level. The intensity of this fluorescence was higher by two orders of magnitude in the tellurite glass because of smaller multiphonon relaxation (see Chapter IVa on Er^{3+}).

Radiative transition probabilities from the emitting ($^5S_2$, $^5F_4$) and $^5F_5$ levels to all terminal levels were calculated using the intensity parameters and matrix elements between the corresponding levels.

The total radiative lifetime τ for $^5S_2$ and $^5F_4$ excited states which are in thermal equilibrium, was calculated using the formula,

$$\frac{1}{\tau_{eff}} = \frac{\sum_a \sum_b g_a \exp(-\Delta E_a/kT)\ A_{ab}}{\sum_a g_a \exp(-\Delta E_a/kT)} \tag{23}$$

where a are the $^5F_4$ and $^5S_2$ emitting levels and b are all terminating levels. ΔE_a is the energy difference between the $^5F_4$ and $^5S_2$ states (50 cm^{-1}).

The branching ratio

$$\beta = \frac{A_{ab}}{\sum A_{ab}} \tag{24}$$

which takes the form in the case of thermalization,

$$\beta = \frac{\sum_a g_a \exp(-\Delta E_a/kT)}{\sum_a \sum_b g_a \exp(-\Delta E_a/kT)\ A_{ab}} \tag{25}$$

Table 16. Calculated branching ratios for emission from excited states of Ho^{3+}

Initial state	Final state	Branching ratio β		
		Tellurite	Calibo	Phosphate
$^5F_5$	$^5I_8$	0.765	0.764	0.798
	$^5I_7$	0.194	0.189	0.198
	$^5I_6$	0.038	0.043	0.042
	$^5I_4$	0.003	0.003	0.003
	$^5I_4$	0.00002	0.00002	0.00002
$^5S_2$	$^5I_8$	0.534	0.538	0.536
	$^5I_7$	0.364	0.367	0.366
	$^5I_6$	0.068	0.063	0.065
	$^5I_5$	0.016	0.015	0.016
	$^5I_4$	0.016	0.015	0.016
	$^5F_5$	0.00002	0.00013	0.00015
$^5F_4$	$^5I_8$	0.780	0.814	0.800
	$^5I_7$	0.108	0.084	0.094
	$^5I_6$	0.071	0.061	0.065
	$^5I_5$	0.031	0.033	0.032
	$^5I_4$	0.005	0.0051	0.005
	$^5F_5$	0.0035	0.0023	0.00254

was also calculated and is presented in Table 16. It is interesting to note that the branching ratios of Ho^{3+} in various glasses and in $YAlO_3$ are very similar, suggesting a similar spectral distribution of the fluorescence in various hosts. This is completely different from Eu^{3+} in which both the branching ratios and spectral distribution of fluorescence are strongly host dependent. Quantum efficiencies of fluorescence from the $^5F_5$ and $^5S_2$ level in tellurite glasses were measured by the comparative method using Tb^{3+} doped glass as a standard (7), and by taking the ratio of the measured lifetime to the effective lifetime. The results obtained by the two methods did not differ by more than 15%. The quantum efficiency of this fluorescence in tellurite glasses is about 10 per cent. As mentioned above, the quantum efficiency in phosphate glasses is about 0.1 per cent due to high multiphonon relaxation.

f) Dysprosium and Praseodymium

The radiative transition probabilities of Dy^{3+} and Pr^{3+} are being calculated at present and will be published soon. The intensity parameters of Dy^{3+} and Pr^{3+} in tellurite glasses are presented in Table 17. The matrix elements U^λ used for

Table 17. Values of τ_λ for Pr^{3+} and Dy^{3+} in tellurite glass

$\tau_\lambda \times 10^9$	Pr^{3+} (14)	Dy^{3+} (16)
τ_2	6.48	20.12
τ_4	15.27	2.74
τ_6	20.98	5.55

calculations of Dy^{3+} were taken from the work of *Carnall et al.* (16) in aqueous solution and for Pr^{3+}, from Ref. (14).

It should be noted that in Pr^{3+}, τ_2 sometimes assumes negative values. A similar result was found by *Weber* for Pr^{3+} in $YAlO_3$ (37). *Weber* suggests that the reason for such behaviour in Pr^{3+} may be due to the proximity of the $5d$ band of Pr^{3+} to $4f$ and of strong f—d admixing (14).

g) Neodymium

Large neodymium glass laser systems capable of delivering multikilojoule sub-nanosecond pulses are being designed at the Lawrence Livermore Laboratory for fusion applications. In this connection, data characterizing the spectral properties of neodymium doped glasses such as: absorption spectra, luminescence emission, radiative lifetimes, quantum efficiencies of fluorescence, induced emission cross sections, excited state absorption are being measured at present by several workers (38, 39, 40). A method for calculating induced — emission cross sections in neodymium laser glasses based on absorption measurements has been demonstrated by *Krupke* (40).

In these calculations, first transition probabilities for emission between excited J manifolds of Nd^{3+} ions in glasses are performed by means of Eq. (13) using experimentally obtained intensity parameters.

The induced — emission cross section σ_p of the $^4F_{3/2} \rightarrow {}^4I_{11/2}$ fluorescence transition at 1.06 μ which is important for the laser design is obtained from the transition probabilities in the following way:

If N is the difference in total population in the $^4F_{3/2}$ and $^4I_{11/2}$ J-manifolds then σ_ϱ, the induced emission cross section gives the maximum spectral growth rate of intensity,

$$I(x) = I_0 \exp (\sigma_\varrho \, Nx) \tag{26}$$

The induced-emission cross section is related to the radiative transition probability by,

$$\sigma_\varrho \, \Delta \, \nu_{\text{eff}} \equiv \int \sigma(\bar{\nu}) \, d\bar{\nu} = \frac{\lambda^2 \varrho}{8 \pi c n^2} \, A(^4F_{3/2}) ; \, (^4I_{11/2}) \tag{27}$$

where $\lambda = 1/\nu$, λ_ϱ is the wavelength of the fluorescence peak and $\Delta\nu_{\text{eff}}$ is the effective fluorescence linewidth determined by numerical integration of the fluorescence line shape (41) and ϱ is the Nd^{3+} ion concentration in the glass.

Intensity parameters have been calculated for a series of Nd^{3+} in alkali alkaline-earth silicate glasses in which 1) the alkali oxide was varied 2) alkaline earth oxide was varied (38). Also the influence of the glass formers on the intensity parameters is studied (38). The data of these parameters are given in Note added in proof and Tables 23—25.

h) Conclusions

It can be seen from the preceding sections that the Judd–Ofelt model is able to predict with satisfactory accuracy (to within 5—10%) the observed electric dipole absorption intensities in a large number of RE in different glasses. It is possible to calculate correctly the radiative transition probabilities and branching ratios of RE in glasses using this model.

On the basis of the results presented above, it is believed that the radiative transition probabilities in many new glasses can be obtained and their validity as lasers predicted. Another factor influencing the radiative quantum yields will be discussed in the following chapter.

V. Nonradiative Relaxations

Nonradiative relaxation between various J states of rare earth ions may occur by the simultaneous emission of several phonons which conserve the energy of the transitions. These multiphonon processes MP arise from the interaction of the electronic levels of the RE with the vibrations of the host lattice. The lattice vibrations are quantized as phonons having excitation energies determined by the masses of the constituent ions and the binding energies between the ions. The MP mechanism by which a non-resonant energy transfer between a donor and an acceptor ion may occur was first formulated by *Orbach* (42) and developed quantitatively by *Miyakawa* and *Dexter* (43) and *Soules* and *Duke* (44). The theory of MP relaxation was further developed by *Fong* and his group (45). *Kiel* (46) treated the problem by applying the time dependent perturbation theory in higher orders. Multiphonon processes involve a combination of crystal field theory and lattice dynamics and because of the tremendous difficulties in the rigorous calculation, suitable approximations are made in all quantitative calculations. The greatest difficulty in a rigorous solution arises from the lack of detailed information regarding the frequency, polarization, symmetry and propagation properties of the vibrations and the strength of the ion-phonon coupling constants. *Riseberg* and *Moos* (47) treated the problem of multiphonon relaxation phenomenologically. This treatment explained successfully the experimental data of relaxation in crystals obtained by *Weber* and other authors (31).

Since the phenomenological model provides a satisfactory description of the experimental results in glasses, as well, we shall summarize the essential factors of this approach below. An extensive description of Riseberg-Moos approach and its application to the experimental data in crystals may be found in the review by *Riseberg* and *Weber* (2).

In what follows we shall summarize the basic assumption for the derivation of the formulae which were used in calculation of multiphonon transition rates in glasses.

The isolated RE ion can interact with its crystal field which varies with time because of lattice vibration. The interaction Hamiltonian (47) may be written as:

$$H = V_0 + \sum_i V_i Q_i + \frac{1}{2} \sum_{i,j} V_{ij} Q_i Q_j + \ldots \tag{28}$$

V_0 is the interaction with the static field. The remaining terms, involving the normal coordinates Q_i represent the interaction with the vibrating lattice.

The expression for $W^{(p)}$ (the transition rate between two electronic levels), involving the emission of p phonons may be obtained either using the pth-order time dependent perturbation theory for the second term in Eq. (28) or the pth term in the Hamiltonian can be used, in the first order perturbation theory.

The phenomenological approach (47) using both mechanisms predicts that the MP transition rate has the form:

$$W^{(p)} = W_0 \prod_i (n_i + 1)^{p_i} \tag{29}$$

Here, p_i is the number of phonons emitted with energy $h\omega$, so that $\sum p_i = p$. W_0 is the transition rate for spontaneous decay at $T = 0K$ and n_i is the Bose-Einstein occupation probability,

$$n_i = [e^{\hbar\omega/kT} - 1]^{-1} \tag{30}$$

The 1 in the parenthesis of (30) represents the spontaneous emission of phonons and n_i, the emission stimulated by thermal phonons. If ΔE is the energy gap between two electronic levels, a number p_i of phonons, of energy $\hbar\omega_i$ is required for conservation of energy, and hence the order of the process is determined by the condition,

$$\sum p_i \hbar\omega = \Delta E \tag{31}$$

The critical feature in the temperature dependence of $W^{(p)}$ is the order of the process, which governs the relaxation. It was found in practice that phonons of single frequency may be inserted in Eq. (31) for transitions in several crystals (43, 44, 45). However, this is not always the case (20, 49).

In practice, one does not observe the MP transition rate between two single discrete levels. In reality, decay occurs between groups of Stark levels of two J multiplets. If levels of the upper multiplet are designated by the subscript a and levels of the lower multiplet by the subscript b, then the total decay rate from one of the upper levels is $\sum_b W_{ab}$. Since the levels within the multiplet are in thermal equilibrium, the combined rate is a Boltzmann average from the separate levels given by,

$$W = \frac{\sum_a \sum_b g_a\, W_{ab}(T)\, \exp(-\Delta E/kT)}{\sum_a g_a\, \exp(-\Delta E/kT)} \tag{32}$$

where W_{ab} is an individual decay rate from upper multiplet level a, to lower multiplet b, g_a is the degeneracy and ΔE is the energy separation of the ath level from the upper level of the upper multiplet. A similar expression exists for the temperature dependence of the total radiative rate, where W_{ab} is replaced by $(W_{ab} + A_{ab})$.

The intrinsic temperature dependence of the multiphonon rate W_{ab} is expressed as follows: (using the single frequency model):

$$W(T) = W_0 \left[\frac{\exp(\hbar\omega/kT)}{\exp(\hbar\omega/kT) - 1} \right]^{p_i} = W_0 \left[1 - \exp\left(-\hbar\omega/kT\right) \right]^{-p_i} \tag{33}$$

The significance of equation (33) is in the possibility of establishing the order of the multiphonon decay process and the energies of the dominant phonons involved.

The dependence of the multiphonon rates on the energy gap, given by *Miyakawa* and *Dexter* (*43*) is of the form:

$$W^{p_i} = W^0 \exp\left(-\alpha \Delta E\right) \tag{34}$$

where W^0 is the transition probability extrapolated to zero energy gap, ΔE is the energy gap between two successive electronic levels and α is expressed by,

$$\alpha = (\hbar\omega)^{-1} \ln[p/g(n+1)]^{-1} \tag{35}$$

where g is the electron phonon coupling constant.

Equation (34) is obtained under assumption that the electron-phonon coupling is weak (which is the case for rare earth ions), and that the occupation number n is not much greater than unity.

The exponential dependence of Eq. (34) is only approximate, because both α and W^0 depend on ΔE.

The treatment of multiphonon relaxation using the adiabatic approximation by *Fong et al.* (*45*) predicts a more complicated dependence of W^p on the energy gap. The treatment by *Englman* and *Jortner* (*48*) also predicts that the non-radiative relaxation rate may not vary exponentially with temperature.

In the phenomenological model the energy gap dependence of the multiphonon transition rate arises as a consequence of the convergence of the perturbation expansion (2). Considering a single frequency model the rate for the pth order process can be written phenomenologically as,

$$W^p = A\omega^p \tag{36}$$

where A is a constant $p = \dfrac{\Delta E}{\hbar \omega}$ and ε is a coupling constant whose magnitude determines the rapidity of convergence of terms in the perturbation expansion.

We see that the multiphonon relaxation probability in all approaches has an approximately exponential dependence on the energy gap between levels.

Furthermore, it was shown by *Miyakawa* and *Dexter* (*43*) that in the case of weak coupling (small g), the relative contribution of higher order multiphonon processes will be more pronounced in optical absorption and emission (phonon side bands) spectra than in relaxation effects.

The validity of the phenomenological approach depends on the extent to which the individual phonon modes and electronic states are statistically averaged out. The critical factor in the multiphonon process is the number of phonons required to conserve energy (the energy gap). Because of the large number of equivalent processes involving many states the symmetry properties of the electronic wavefunction are generally unimportant. The phonon frequency ω_i should be close to the maximum frequency since such a process will involve the smallest number of phonons and smallest possible order of interaction. This is usually true in crystals and the experimental studies of the temperature dependence confirm that the order of the process is often the lowest consistent with energy conservation and the cut off frequency of the phonon spectrum. If lower frequency modes are more numerous and strongly coupled, the dominant multiphonon process may occur in higher order. Recently, *Reed* and *Moos* (*49*) have found that in multiphonon relaxation of RE in YVO_3, $YAlO_3$ and YPO_4 low frequency phonons are also involved in the multiphonon process in addition to high energy phonons.

The mechanism of multiphonon relaxation of RE ions in glasses was studied during the last two years and the results of these studies are presented below.

Multiphonon relaxation of Eu^{3+}, Er^{3+} and Tm^{3+} in borate, phosphate, germanate and tellurite glasses at room temperature was studied by *Reisfeld et al.* (*50*). The purpose of this study was to predict and control quantum efficiencies from desired levels of the rare earth ions.

This idea is first explained for Eu^{3+} doped glasses.

a) Europium

Figure 9 presents part of the emission spectra of Eu^{3+} in various glasses. The transitions appearing in the figure are from $^5D_3$, $^5D_2$ and $^5D_1$ excited levels to the ^7F manifold of the ground state. The emission from the $^5D_0$ level to the ^7F manifold is by one order of magnitude higher and is not presented in this figure. The assignment of the bands appearing in Fig. 9 are given in Table 18. As can be seen from this figure, the transitions from $^5D_3$ and $^5D_2$ to ^7F manifold do not appear in borax and phosphate glasses. A notable increase in the emission intensity is observed in the order of borax $<$ phosphate $<$ germanate $<$ tellurite. The relative areas under the fluorescence curves of $^5D_1$ and $^5D_0$ are given in Table 19.

The decay times of the fluorescence from the $^5D_0$ level of Eu^{3+} were simple exponentials in all glasses. Their values are 2.83 msec in phosphate, 1.7 msec in germanate and 1.01 msec in tellurite.

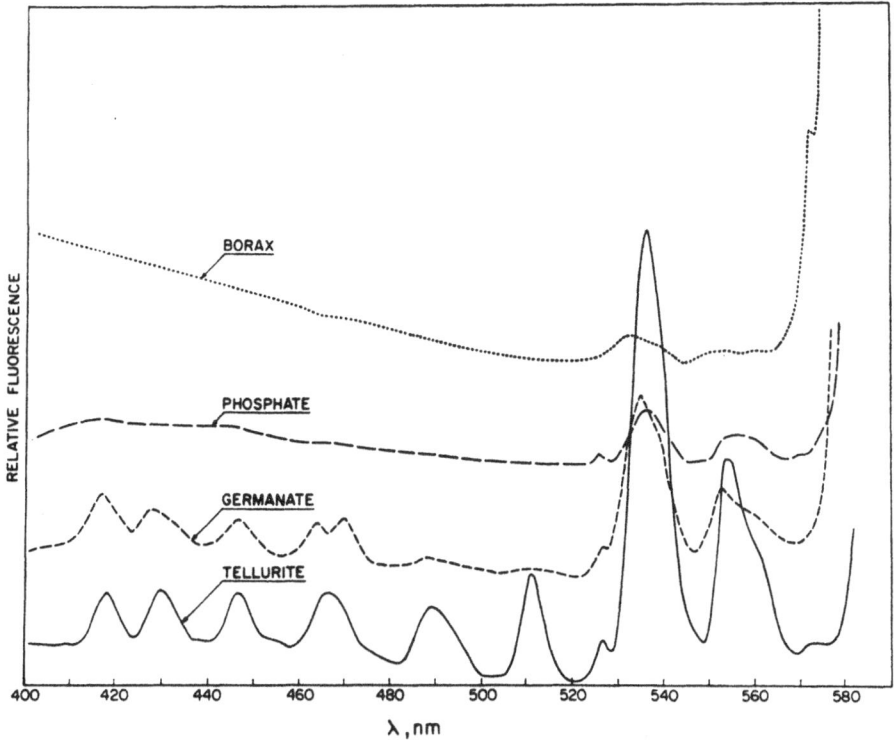

Fig. 9. Part of the emission spectra of Eu^{3+} in various glasses

Table 18. Assignments of the observed fluorescence bands of Eu^{3+} [a])

Band	Assignment
416 nm	$^5D_3 \rightarrow {}^7F_0$
430 nm, 448 nm	not defined
464 nm	$^5D_2 \rightarrow {}^7F_0$
470 nm	$^5D_2 \rightarrow {}^7F_1$
490 nm	$^5D_2 \rightarrow {}^7F_2$
510 nm	$^5D_2 \rightarrow {}^7F_3$
527 nm	$^5D_1 \rightarrow {}^7F_0$
536 nm	$^5D_1 \rightarrow {}^7F_1$
555 nm	$^5D_1 \rightarrow {}^7F_2$

[a]) The transitions from $^5D_0$ are not presented.

Table 19. Relative areas of the emission of Eu^{3+} in various glasses

Transition	Borate	Phosphate	Germanate	Tellurite
$^5D_0 \to {}^7F_0$	100	100	100	100
$^5D_1 \to {}^7F_2$	—	14.5	24.9	82.3
$^5D_1 \to {}^7F_1$	2.5	21.3	35.0	176.2
$^5D_1 \to {}^7F_0$	0.9	2.0	0.7	3.8

The transition probability K_{10} from state $^5D_1$ to $^5D_0$ of Eu^{3+} was calculated using the following rate equations for depopulation of the system.

$$\frac{dN_1}{dt} = - (k_{10} + k_{1g}) N_1 \tag{37}$$

$$\frac{dN_0}{dt} = k_{10} N_1 - (k_{0g}^r + k_{0g}^{nr}) N_0 \tag{38}$$

The symbols in these equations have the following meanings: k_{1g} — total transition probability from the $^5D_1$ level to the ^7F ground multiplet; k_{0g} — transition probability from the $^5D_0$ level to the ^7F multiplet; "r" and "nr" — radiative and nonradiative transitions respectively; N_1 and N_0 — the population of the levels $^5D_1$ and $^5D_0$ at a given time; $k_{0g}^{nr} \to 0$ because of the high energy gap between the $^5D_0$ and ^7F levels.

Let
$$J_0 = \int N_0(t)\, dt$$
$$J_1 = \int N_i(t)\, dt \tag{39}$$

The time integrated luminescence intensity I of a state $|j>$ of the system is then given by:

$$I_j = K_j^r J_j \tag{40}$$

which holds for continuous irradiation. K_j^r is the radiative transition probability from the j-th level ($^5D_1$ or $^5D_0$) to the ground ^7F multiplet.

Integration of Eq. (37) and (38) using Eq. (39) gives:

$$- N_1(0) = - (k_{10} + k_{1g}) J_1 \tag{41}$$

$$- N_0(0) = k_{10} J_1 - k_{0g}^r J_0 \tag{42}$$

If the absorption is into level $|^5D_1>$ then $N_0(0) = 0$.

$$k_{10} J_1 = k_{0g}^r J_0 \tag{43}$$

From Eq. (31),

$$J_1 = \frac{I_1}{k_{1g}^r} \quad \text{and} \quad J_0 = \frac{I_0}{k_{0g}^r}$$

Then, Eq. (43) becomes:

$$k_{10} \frac{I_1}{k_{1g}^r} = I_0 \tag{44}$$

from which it follows that:

$$k_{10} = \frac{I_0}{I_1} k_{1g}^r \tag{45}$$

The numerical value of k_{1g}^r can be calculated from the transition probability $^7F_0 \to {}^5D_1$ which was obtained from absorption and the relative areas (39) of all the transitions $^5D_1 \to {}^7F_i$

$$k_{1g}^r = \frac{1}{\sum \tau_{nat}} = k(^5D_1 \to {}^7F_0) \left[1 + \frac{\sum_{i=1}^{6} S(^5D_1 \to {}^7F_i)}{S(^5D_1 \to {}^7F_0)} \right] \tag{46}$$

where S is the area under the relevant transition. It is assumed in Eq. (46) that $k(^7F_0 \to {}^5D_1) = k(^5D_1 \to {}^7F_0)$. The value $\frac{1}{\tau_{nat}} = k(^7F_0 \to {}^5D_1)$ is taken from formula (47), which is valid for narrow level system.

$$\frac{1}{\tau_{nat}} = 2.88 \times 10^{-9} n^2 \frac{gl}{gu} <\nu>^2 \int \varepsilon(\nu) \, d\nu \tag{47}$$

where n is the refraction index, $<\nu>^2$ the squared average wavenumber of the absorption, gl, gu, the degeneracies of the lower and upper states, $\varepsilon(\nu)$ the extinction coefficient as a function of wavenumber.

The values of k_{1g}^r were calculated using Eq. (46) in which only the transitions $^5D_1 \to {}^7F_{0,1,2}$ were included. Other transitions were hidden under the emission bands of the $^5D_0 \to {}^7F$ multiplet.

The values so obtained for k_{1g}^r were 53.7 sec^{-1} in phosphate, 70.48 sec^{-1} in germanate, and 57.05 sec^{-1} in tellurite. From these, the transition probabilities k_{10} were obtained: 1.57 10^4 sec^{-1}, 7.05 10^3 sec^{-1} and 3.98 10^3 sec^{-1} for phosphate, germanate and tellurite respectively[1].

Weber (17) in his work on Eu^{3+} doped LaF$_3$, calculated the radiative probabilities for all the transitions from $^5D_1$ to ^7F multiplet and obtained the ratio,

$$\frac{k(^5D_1 \to {}^7F_{0,1,2,3,4,5,6})}{k(^5D_1 \to {}^7F_{0,1,2})} = \frac{130}{80}.$$

---

[1] Calculations of radiative transition probabilities in Eu^{3+} are being performed in this laboratory (57).

It may be expected that a similar ratio will be obtained in the case of glasses. Therefore, the actual value of k_{1g}^r should be greater by such a factor, increasing correspondingly the value of k_{10}.

b) Thulium

The electronic transitions $^1G_4 \rightarrow {}^3H_4$ ($\sim$652 nm) and $^3F_2 \rightarrow {}^3H_6$ ($\sim$670 nm) of Tm^{3+} in various glasses are presented in Fig. 10. The excitation was to the $^1G_4$ level and the emission spectra were corrected for the absorption of this level in the different media. Thus, this figure presents the relative quantum efficiencies.

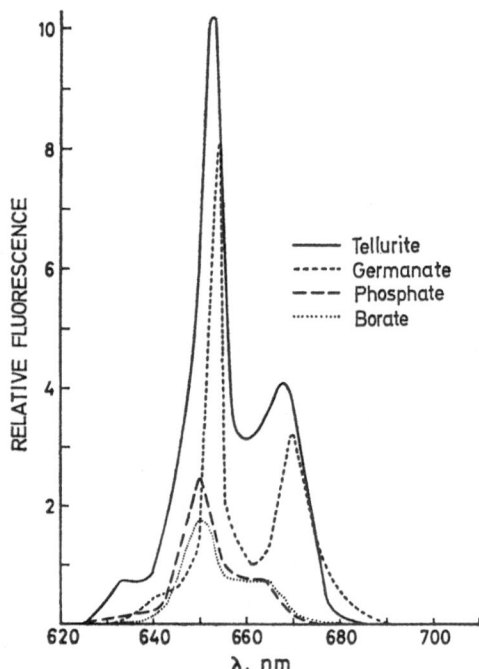

Fig. 10. Emission spectra of Tm^{3+} in various glasses normalized for the absorption Excited at $^1G_4$ (469 nm)

It is evident from the figure that the emission intensities increase in order similar to that observed in Eu^{3+}. Using the experimentally obtained emission intensities from the $^1G_4$ level and the absorption coefficients, the probability of relaxation from the higher $^1D_2$ level to the next lower level $^1G_4$ was calculated. The rate equations for such cases are similar to those developed by *Nakazawa* and *Shionoya* (52) and the present authors for depopulation of $^1D_2$ to $^1G_4$ emitting levels of

Tm^{3+} (53). The final equation used for calculation of the relaxation rate K_{24} from $^1D_2$ to $^1G_4$ is

$$K_{24} \frac{I_{24} J_4}{I_{44} J_2} \cdot \frac{B_4}{B_2} \cdot \frac{1}{\tau_2} \tag{48}$$

Here, I represents the emission intensities, and the suffixes 2 and 4 mean the levels $^1D_2$ and $^1G_4$. The first suffix indicates the level excited and the second one the emitting level. B and J denote the absorption cross-section and the exciting photon flux, respectively. τ_2 is the measured lifetime of the $^1D_2$ level. The values of K_{24} thus obtained for various glasses are presented in Table 20.

c) Erbium

Fig. 11 presents the relative spectral efficiencies of the $^2H_{9/2} \rightarrow \,^4I_{15/2}$ (~ 408 nm), $^2H_{11/2} \rightarrow \,^4I_{15/2}$ (~ 525 nm), $^4S_{3/2} \rightarrow \,^4I_{15/2}$ (~ 535, 546, 556 nm) and $^4F_{9/2} \rightarrow \,^4I_{15/2}$ (~ 660 nm) transitions of Er^{3+} in germanate and tellurite glasses. The spectra presented in this figure are corrected for the absorption of the $^4G_{11/2}$ (~ 380 nm)

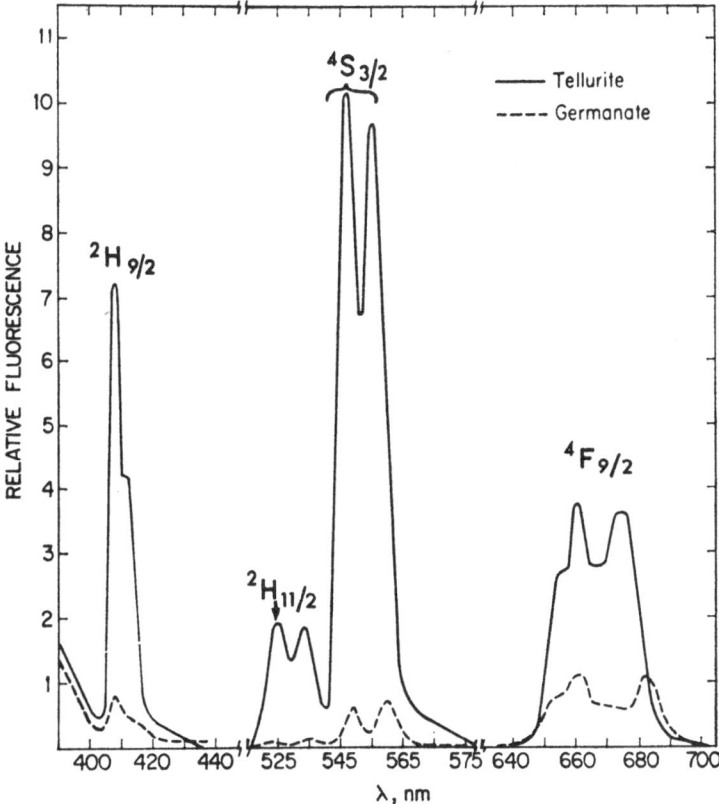

Fig. 11. The emission spectrum of Er^{3+} in germanate and tellurite glasses normalized for the absorption. Excited at $^4G_{11/2}$ (379 nm)

Table 20. Comparison of multiphonon relaxation rates, W, from $^1D_2$ to $^1G_4$ levels of Tm^{3+}, $^2H_{11/2}$, $^4S_{3/2}$ to $^4F_{9/2}$ and $^4F_{9/2}$ to $^4I_{9/2}$ levels of Er^{3+} in various glasses by use of different experimental data

Rare earth ion	Equation used for the calculation	W of transitions in:					
		Borate	Phosphate	Germanate		Tellurite	
		80 K	80 K	80 K	300 K	80 K	300 K
a)Tm^{3+}	$W = \dfrac{1}{\tau_{meas}}(D) - \sum A_{ed}(D)$	3.6×10^4	1.9×10^4	1.5×10^4		3.9×10^4	
	$W = \sum A_{ed}(D)\,\dfrac{1-\eta(D)}{\eta(D)}$	3.6×10^4	2.0×10^4	1.54×10^4		3.9×10^4	
Er^{3+}	$W = \dfrac{1}{\tau_{meas}}(H, S) - \sum A_{eff}(H, S)$			0.73×10^5	1.06×10^5	0.89×10^4	1.34×10^4
	$W = \sum A_{ed}(S) - \dfrac{1-\eta(S)}{\eta(S)}$			0.75×10^5	1.36×10^5	0.83×10^4	1.07×10^4
	$W = \sum A_{ed}(F) - \dfrac{1-\eta(F)}{\eta(F)}$			1.88×10^5	2.85×10^5	1.56×10^5	2.33×10^5

a) The nonradiative rates of Tm^{3+} obtained from the experimental data in which the measured lifetimes of the $^1D_2$ level of Tm^{3+} and the radiative lifetime provide a value of 0.65 for quantum efficiency seems to be too high. A quantum efficiency of 0.99 which is much more probable would give $W \sim 6 D 10^2$.

excited level in the two different media and corrected for the spectral distribution of the light source, response of the photomultiplier and monochromators. As can be seen, the fluorescent quantum efficiency in tellurite is much higher than in germanate glass. No fluorescence was observed for Er^{3+} doped borate glass and only very weak fluorescence was obtained for Er^{3+} in phosphate glass.

d) Temperature Dependence of Erbium Fluorescence

Fig. 12 presents the fluorescence spectra of Er^{3+} in tellurite at 80 K and 300 K in the range of 520—560 nm and of 650—680 nm under $\sim$380 excitation. The increase at low temperature of the $\sim$550 and the $\sim$660 nm emissions in tellurite is slightly higher than in the germanate and is by a factor of about 2 in respect to room temperature intensities. The peaks at $\sim$525 nm and $\sim$535 nm originate from rapid thermalization of the $^{2}H_{11/2}$ and the states of the $^{4}S_{3/2}$ multiplets. These peaks disappear in the spectra of Er^{3+} in 80 K, but they grow in intensity with increase of temperature, while the intensity of the 550—560 nm peaks decreases. Figs. 13 and 14 present the dependence of the Er^{3+} ($^{2}H_{11/2}$, $^{4}S_{3/2}$) $\rightarrow$ $^{4}I_{15/2}$ emissions in tellurite and germanate glasses, respectively. In the tellurite glasses the ($^{2}H_{11/2}$, $^{4}S_{3/2}$) $\rightarrow$ $^{4}I_{15/2}$ emissions show a one step decrease above 140 K due to ($^{2}H_{11/2}$, $^{4}S_{3/2}$) $\rightarrow$ $^{4}I_{15/2}$ quenching, while in the germanate glasses, a two-step decrease is observed: the mild decrease above 320 K in the germanate

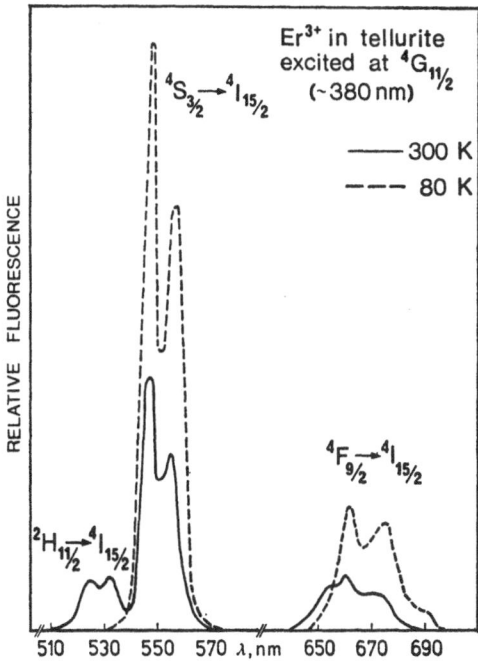

Fig. 12. Fluorescence spectra of Er^{3+} in tellurite glass at 80 K and 300 K due to $^{2}H_{11/2}$, $^{4}S_{3/2} \rightarrow {}^{4}I_{15/2}$ ($\sim$520—560 nm) and due to $^{4}F_{9/2} \rightarrow {}^{4}I_{15/2}$ ($\sim$650—680 nm) emissions at $^{4}G_{11/2}$ ($\sim$380 nm) excitation

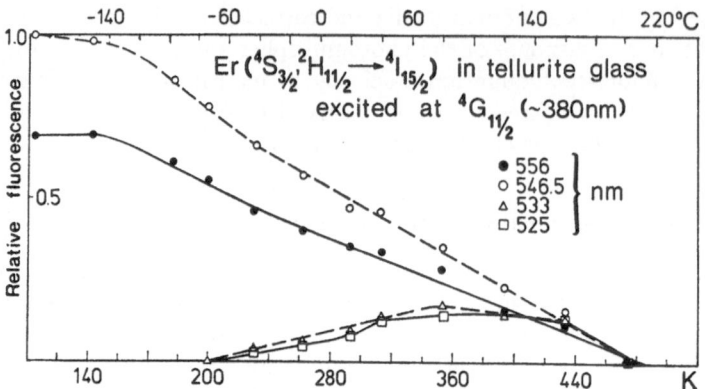

Fig. 13. Dependence of $^2H_{11/2}$, $^4S_{3/2} \to {}^4I_{15/2}$ emission of Er^{3+} on temperature in tellurite glass

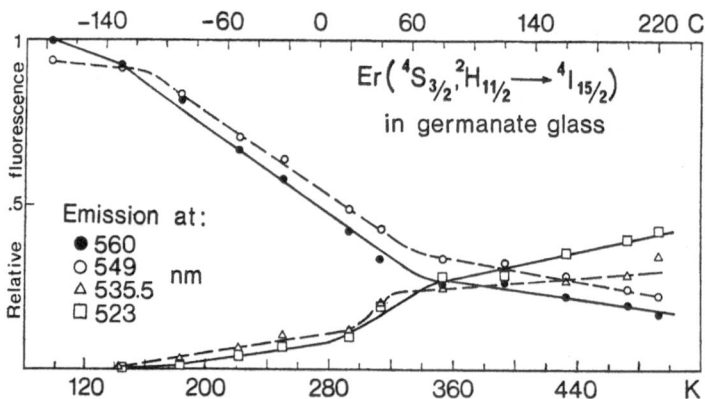

Fig. 14. Dependence of $^2H_{11/2}$, $^4S_{3/2} \to {}^4I_{15/2}$ emission of Er^{3+} on temperature in germanate glass

glasses can be ascribed to the increased transparency of the glass host, permitting more rare earth centers to be excited, as in the case of Tm^{3+} doped glasses.

In all the calculations of W from the Er^{3+} $^4S_{3/2}$ level the effect of thermalization of the $^2H_{11/2}$ state must be taken into account. Since the radiative probabilities of $^2H_{11/2}$ are significantly different than that of $^4S_{3/2}$ (29), assuming rapid thermalization, the radiative effective decay time, τ_{eff}, will be given by:

$$\frac{1}{\tau_{eff}} = \sum A_{eff}{}^R = \sum_{[i = {}^2H_{11/2'},\ {}^4S_{3/2}]} g_i A_i \{e^{-(E_1-E_i)/kT}\} / \sum g_i \{e^{-(E_1-E_i)/kT}\} \qquad (49)$$

where A_i is the total spontaneous emission probability from level i to all lower levels and g_i is the level degeneracy.

The values of W can be obtained for various temperatures from the temperature dependence of fluorescence intensities and its known value at one temperature from the following relation:

$$I_1 = I_0 \, \eta \, (T_1) = I_0 \, \frac{\sum A_{ij}}{\sum A_{ij} + \sum W_1} \tag{50}$$

where I_0 is proportional to the number of initially excited centers, I_1 is the intensity at temperature T_1 and $\sum W_1$ is the total nonradiative relaxation rate at temperature T_1. Since $\sum A_{ij}$ is temperature independent, from the ratio of I_n to I_1 we obtain:

$$\sum W_n = \frac{I_1}{I_n} \left(\sum A_{ij} + \sum W_1 \right) - \sum A_{ij} \tag{51}$$

where I_n is intensity at temperature T_n.

In the event that $\sum A_{ij}$ does depend on temperature because of thermalization of higher states with different radiative transition rates, as in the case of the Er^{3+} $^4S_{3/2}$ level, $\sum A_{ij}$ at a temperature T_1 is given by Eq. (49), and Eq. (50) becomes:

$$I_1 = I_0 \left(\frac{I_i}{\beta_i} + \frac{I_j}{\beta_j} \right) T_1 = I_0 \, \eta_{total} \, (T_1) = I_0 \left(\frac{\tau_{mes}}{\tau_{eff}} \right) T_1 = I_0 \frac{\sum A_{eff} \, (T_1)}{\sum A_{eff} \, (T_1) + \sum W_{eff} \, (T_1)} \tag{52}$$

Here, $I_1 = \left(\frac{I_i}{\beta_i} + \frac{I_j}{\beta_j} \right)_1$ is the total emission intensity at temperature T_1, I_i is the emission intensity from the excited ($^4S_{3/2}$ state in case of Er^{3+}) and I_j is the emission from the thermalized level j ($^2H_{11/2}$ state in case of Er^{3+}). The β_i and β_j are the branching ratios of the measured transitions from i and j levels, as defined by Eq. (50) and Table 13. Now, the Eq. (51) takes a form:

$$\sum W_{eff} \, (T_n) = \frac{I_1}{I_n} \sum A_{eff} \, (T_n) \left[1 + \frac{\sum W_{eff} \, (T_1)}{\sum A_{eff} \, (T_1)} \right] - \sum A_{eff} \, (T_n) \tag{53}$$

Since the measurements of intensities are more accurate with our experimental set up than the life time measurements, the temperature dependence of lifetimes was studied only in the case of Er^{3+} in tellurite glass while in the remaining cases the temperature dependence of W was obtained from experimentally measured temperature dependent fluorescent intensities using Eq. (51) or (53).

e) Temperature Dependence of MP Relaxation Rates

When only multiphonon losses are responsible for the temperature dependence of intensities, the dependence of W with temperature follows Eq. (30). In order to test this, ln (W) was plotted versus $y \equiv \ln(1 - e^{-\hbar\omega/kT})$ in Fig. 15. A straight line was obtained for W vs. y in case of the Er^{3+} $^4S_{3/2}$ excited level in tellurite glasses. The slope of this corresponds to 4 phonons, namely an energy gap of

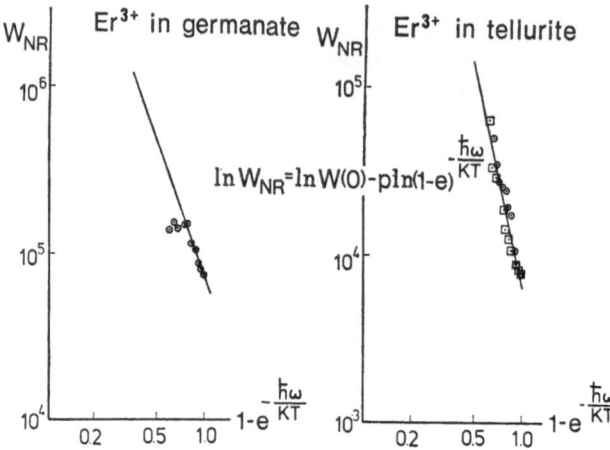

Fig. 15. Plot of ln W versus ln $(1 - e^{-\hbar\omega/kT})$ of Er^{3+} $^4S_{3/2}$ excited level in germanate and tellurite glasses.
⊙ — calculated from the measured lifetimes of $^2H_{11/2}$, $^4S_{3/2}$ levels;
☐ — calculated from the measured emission intensities from $^2H_{11/2}$, $^4S_{3/2}$ levels

2700 cm^{-1}, assuming that the wavenumber of the high energy phonons is 675 cm^{-1} consistent with the lowest order of relaxation. The energy gap between the baricenters of $^4S_{3/2}$ and $^4F_{9/2}$ levels is 2930 cm^{-1}. Our experimental results correspond fairly well to the 4 phonon process. In germanate a straight line was obtained up to 313 K; at higher temperatures a deviation from linearity was observed. The slope corresponds to 3 phonons of GeO_2 stretching frequencies matching the energy gaps. From these results for the $^4S_{3/2}$ excited level of Er^{3+}, we see that the MP relaxation is responsible for the nonradiative losses. It is therefore concluded that the lowest order of multiphonon relaxation in which the most energetic phonons are involved is effective in glasses.

The situation in Tm^{3+} is much more complicated as additional factors such as energy transfer from tellurite and borate to Tm^{3+} and absorption of emitted light by the tellurite were observed. Because of these complications we were not able to extract experimentally the contribution of the multiphonon relaxation to the temperature dependence of fluorescence in Tm^{3+}.

f) Dependence of Nonradiative Transition Probabilities on Energy Gap and Host Matrix

The logarithmic MP relaxation rates, W, for Tm^{3+} $^1D_2 \rightarrow {}^1G_4$, Er^{3+} $^4S_{3/2} \rightarrow {}^4F_{9/2}$ and $^4F_{9/2} \rightarrow {}^4I_{9/2}$ transitions in different glass matrices, measured at 80 K are plotted versus energy gap ΔE and versus normalized energy gap $\equiv \dfrac{\Delta E}{\hbar\omega}$ (where $\hbar\omega$ is the energy of the highest-energy glass phonons) in Fig. 16. In this figure we inserted also the values of spontaneous multiphonon relaxation rates for Er^{3+} $^4S_{3/2} \rightarrow {}^4F_{9/2}$ and $^4F_{9/2} \rightarrow {}^4I_{9/2}$ transitions in MnF_2 calculated by *Flaherty*

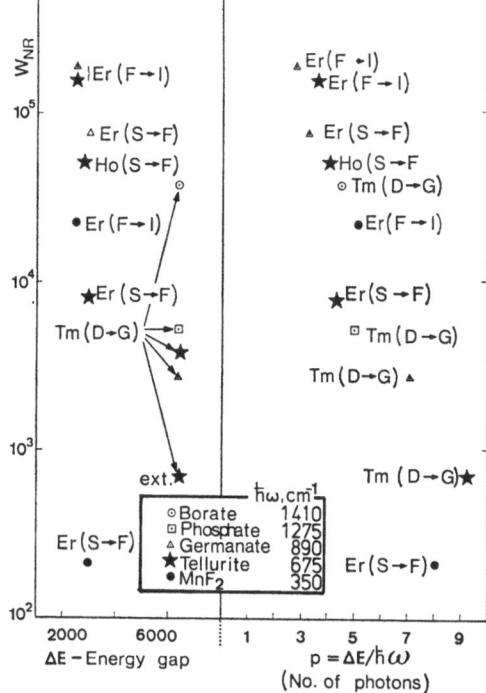

Fig. 16. Logarithmic plot of multiphonon relaxation rates, W, for different rare earths in various host matrices versus energy gap, ΔE, and versus number of phonons matching the energy gap

and *DiBartolo* (*54*). From Fig. 14, it is evident that the W $\approx$ W$_{0K}(\Delta E)$ values decrease as the energy gap increases. For a given energy gap, the most critical factor affecting the multiphonon relaxation is the energy of the host lattice phonons: the lower the phonon energy, the smaller are the nonradiative rates obtained. In addition, since almost a straight line is obtained when the W values for various hosts are plotted versus $\dfrac{\Delta E}{\hbar\omega}$ $\equiv$ number of phonons matching a given energy gap, we conclude that the nonradiative relaxation is dominated by the number of phonons needed to match the energy gap. Were there any changes in the orbit-lattice coupling, they would be insignificant for the systems studied and within the experimental error. This result is consistent with the single frequency model and with the assumption that the multiphonon relaxation in glass is of the lowest order in which the contribution of the most energetic phonons is dominant. Hence, Eq. (54) represents the physical picture:

$$W(\Delta E) = W(0)\, e^{-p}. \tag{54}$$

It should be noted that the accuracies of the calculated nonradiative probabilities are dependent upon the experimental uncertainties in the lifetime measurements ($+10\%$) and the error in evaluating the radiative transition probabilities.

It was proposed earlier (1) that in oxide glasses, a rare earth ion is surrounded by eight non-bridging oxygens belonging to the corners of XO_4 (X = B, P, Ge) glass forming tetrahedra or octahedra of TeO_2, each polyhedron donating two oxygens. As a result, a distorted cube is formed surrounding the RE ion.

In rare earth doped glasses there are two main groups of phonons: the high energy phonons arising from the X—O stretching frequencies and the low energy phonons arising from the Me—O bond (Me = RE^{3+}, Na^+, K^+, etc.). In general, the highest energy phonons are considered to make the dominant contribution to the nonradiative relaxation, as the multiphonon relaxations occur by emitting the smallest number of phonons which match the energy gap between two successive electronic levels. Therefore, the X—O stretching frequencies with energies, as given in Table 21, are mainly responsible for the phonon assisted

Table 21. Phonon energies of various oxidic glasses

Glass	Bond	Phonon Energy (cm^{-1})
Borate	B—O	1340—1480
Phosphate	P—O	1200—1350
Germanate	Ge—O	975—800
Tellurite	Te—O	750—600

relaxation in glasses. The low energy phonons will become effective if the gap is not exactly matched by the higher energy phonons. From our experimental results, we have concluded that the variation in the electron phonon coupling constant g in various glasses would not explain the observed enhancement in the emission intensities. As mentioned above, the relation between the radiative and nonradiative probabilities and quantum efficiencies η of fluorescence is given by

$$\eta = \frac{\Sigma A}{\Sigma A + \Sigma W} \tag{55}$$

where A represents the radiative and W the nonradiative transition rates.

As seen from the formula, a decrease in the nonradiative probabilities in cases where the radiative constants are similar, results in a higher quantum efficiency. Hence η is governed by the energies of the phonons of the host matrix.

In Eu^{3+}, the energy gaps between the metastable levels from which fluorescence may be observed are:

$$^5D_0 \rightarrow {}^7F_0 \approx 17{,}000 \text{ cm}^{-1}$$

$$^5D_1 \rightarrow {}^5D_0 \approx 1{,}650 \text{ cm}^{-1}$$

$$^5D_2 \rightarrow {}^5D_0 \approx 2{,}600 \text{ cm}^{-1}.$$

In phosphate and borate glasses, where one phonon of the X—O bond is needed for the matching of the energy gap between $^5D_1 \rightarrow {}^5D_0$, the nonradiative rate is high and the fluorescence from $^5D_1$ level is very low. In germanate and tellurite glasses, the energy gap can be matched by 2—3 X—O phonons and fluorescence from $^5D_1$ and even $^5D_2$ is observed with high intensity.

In erbium, the energy gaps between two close lying electronic levels are about 1 000—3 000 cm^{-1}. As a consequence of the small energy gaps combined with the high phonon energies of borate and phosphate, rapid non-radiative depopulation of the erbium levels situated between 32,000 and 10,000 cm^{-1} occurs, resulting in a very low emission intensity. In germanate and tellurite, which have smaller phonon energies than borate and phosphate, the nonradiative losses are smaller and intense fluorescence from $^2H_{9/2}$, $^2H_{11/2}$, $^4S_{3/2}$ and $^4F_{9/2}$ is observed.

In thulium, as in erbium, the emission intensities are the highest in the tellurite glasses. The energy gaps between the various electronic levels of thulium situated between 35,000 and 15,000 cm^{-1} are about 6,000 cm^{-1}, which correspond to the energy of 4 phonons in borate, about 5 in phosphate, 7 in germanate and 9—10 in tellurite glasses. This is consistent with the fact that the observed fluorescence is the highest in tellurite.

From our experimental results, we see that the dependence of emission intensities on the phonon energies is stronger when the energy gap, ΔE, between the emitting levels is smaller. For example, in the case of emission from Er $^4S_{3/2}$ (ΔE ($^4S_{3/2} - {}^4F_{9/2}$) $\sim$2650 cm^{-1}), the fluorescence is higher by a factor of 15 in the tellurite than in the germanate glass, while in the case of emission from Tm^{3+} $^1G_4$ (ΔE ($^1G_4 - {}^3F_2$) $\sim$6000 cm^{-1}), the ratio of fluorescence in tellurite to germanate is only 1.3.

The numerical values of nonradiative rate constant differ much less than the values of quantum efficiencies in various glasses. As seen from formula (2) the quantum efficiency depends on both the nonradiative and the radiative transition rates. Therefore, the increase in the quantum efficiency is not always followed by a similar increase in the multiphonon relaxation rate.

In summary, we can see that by incorporating the activator ion in the glass host of a proper phonon energy, it is possible to increase the quantum efficiency of emission and control the levels from which emission occurs.

g) Fluorescence of RE in Beryllium Fluoride Glasses (FBG)

The theory of multiphonon relaxation in glasses developed in the proceeding section can be tested also in nonoxide glasses such as BeF$_2$ (with admixtures of KF, CaF$_2$ and AlF$_3$) (FBG). Luminescence of RE in BeF$_2$ was summarized in a review by *Kolobkov* and *Petrovskiy* (55). The most outstanding feature in these systems is that additional fluorescence bands to those observed in silicate and phosphate glass are observed. Strong fluorescence is observed from the $^5D_1$ level of Eu^{3+} similarly to the fluorescence observed in tellurite glasses. Also, relative fluorescence yield from the $^5D_3$ level of Tb^{3+} to that of $^5D_4$ is high in [Ref. (56)], FBG.

Ho^{3+} in FBG exhibits visible fluorescence similarly to tellurite glasses. No quantitative calculation can be performed of the nonradiative relaxation rate;

Table 22. Relative areas and half bandwidths of the $^4G_{5/2}$ emission of Sm^{3+} excited at 402 nm

Transition assignment	Wavelength (nm)	1/2 B.W. cm^{-1}			R.			R.A.[a]		
		Borate glass	Borax glass	GeO$_2$ glass	TGG	Borate glass	Borax glass	GeO$_2$ glass	TGG	Borate glass	Borax glass	GeO$_2$ glass	TGG
$^4G_{5/2} \rightarrow {}^6H_{5/2}$	~565	388	402	248	355	1	1	1	1	25.0	27.3	1	30
$^4G_{5/2} \rightarrow {}^6H_{7/2}$	~602	502	542	205	444	2.76	2.70	3	2.38	72.3	74.5	3	69.7
$^4G_{5/2} \rightarrow {}^6H_{9/2}$	~649	214	230	107	184	2.1	2.0	3.5	2.69	56.0	58.3	3.5	77.6

[a] Relative to $^4G_{5/2} \rightarrow {}^6H_{5/2}$ in GeO$_2$.

FBG, because of lack of data of the absolute absorption. However qualitatively it can be seen that the existence of appreciable fluorescence from higher levels in the visible fluorescence indicates that nonradiative relaxation rates are smaller than in silicate glasses and similar with tellurite glasses. Since the phonon energies of FBG are around 600 cm^{-1} [Ref. *(56)*] this result is not surprising.

In conclusion, we see that the highest energy phonons are responsible for the nonradiative relaxation in glasses. This is in accordance with the MP relaxation theories, as described in the previous sections. By incorporating the activator ion into the glass host having a suitable phonon energy it is possible to increase the quantum efficiency of emission and control the levels from which emission occurs.

Note added in proof. Extensive work on the influence of network formers and network modifiers on the intensity parameters of Nd^{3+} in various glasses is being performed at present by Dr. *M. J. Weber* and his colleagues.[2] The following tables present the various parameters of Nd^{3+} as a function of glass network former and modifier. The author is deeply grateful to Dr. *Weber* for providing the tables.

Table 23. Intensity parameters of Nd^{3+} in alkali alkaline-earth silicate glasses

Glass composition (mole %)			
SiO$_2$	65	65	65
BaO	20	20	20
Li$_2$O	15		
Na$_2$O		15	
K$_2$O			15
Nd^{3+} intensity parameters (10^{-20} cm^2)			
Ω_2	3.8	3.8	3.7
Ω_4	3.4	3.4	2.8
Ω_6	4.1	3.5	2.3

Table 24. Intensity parameters of Nd^{3+} in alkali alkaline-earth silicate glasses

Glass composition (mole %)				
SiO$_2$	65	65	65	65
K$_2$O	15	15	15	15
MgO	20			
CaO		20		
SrO			20	
BaO				20

[2] *Layne C. B., Lowdermilk W. H., Weber M. J.:* to be published.

Table 24 (continued)

Nd³⁺ intensity parameters (10^{-20} cm²)				
Ω_2	5.1	4.6	3.8	3.7
Ω_4	4.1	3.6	3.3	2.8
Ω_6	2.8	2.9	2.7	2.3

Table 25. Intensity parameters of Nd^{3+} in alkali alkaline-earth glasses with 67 mole % of network formers

Nd³⁺ intensity parameters (10^{-20} cm²)				
Glass:	Borate	Silicate	Phosphate	Germanate
Ω_2	4.1	4.0	2.7	5.8
Ω_4	3.4	3.3	5.3	3.3
Ω_6	4.3	2.5	5.4	2.9

Acknowledgement. The author gratefully acknowledges the benefit of the numerous and helpful discussions with Professors *B. Englman, F. K. Fong, Christian Jørgensen, R. Orbach* and *M. J. Weber*, prior to writing this review.

Im am also grateful to my colleagues and students, *E. Greenberg, B. Barnett, L. Boehm. A. Bornstein, Y. Eckstein, J. Hormodaly* and *N. Lieblich*, who provided me with their help and encouragement during the writing of this manuscript.

VI. References

1. *Reisfeld, R.:* Struct. Bonding *13*, 53 (1973).
2. *Riseberg, L. A., Weber, M. J.:* Relaxation phenomena in rare earth luminescence. In: Progress in Optics, Vol. XIV (ed. *E. Wolf*). Amsterdam: North Holland Publishing Co. 1975, in press.
3. *Orbach, R.:* In: Optical properties of solids (ed. *R. DiBartolo*). Plenum Press 1975.
4. *Watts, R. K.:* In: Optical properties of solids (ed. *R. DiBartolo*). Plenum Press 1975.
5. *Judd, B. R.:* Phys. Rev. *127*, 750 (1962).
6. *Ofelt, G. S.:* J. Chem. Phys. *37*, 511 (1962).
7. *Reisfeld, R.:* J. Res. Natl. Bur. St., A. Phys. Chem. *76A*, 613 (1972).
8. *Wybourne, B. G.:* Spectroscopic properties of rare earths. New York: Interscience 1965.
9. *Carnall, W. T., Fields, P. R., Rajnak, K.:* J. Chem. Phys. *49*, 4412 (1968).
10. *Jørgensen, C. K., Judd, B. R.:* Mol. Phys. *8*, 281 (1964).
11. *Nielson, C. W., Koster, G. F.:* Spectroscopic coefficients for the p^n, d^n and f^n configurations. Cambridge, Mass.: M. I. T. Press 1964.
12. *Carnall, W. T., Fields, P. R., Wybourne, B. G.:* J. Chem. Phys. *42*, 3797 (1965).
13. *Reisfeld, R., Boehm, L., Lieblich, N., Barnett, B.:* Proc. Tenth Rare Earth Conf. *2*, 1142 (1973).
14. *Weber, M. J.:* J. Chem. Phys. *49*, 4774 (1968).
15. *Krupke, W. F., Gruber, J. B.:* Phys. Rev. A *139*, 2008 (1965).
16. *Carnall, W. T., Fields, P. R., Rajnak, K.:* J. Chem. Phys. *49*, 4450 (1968).
17. *Weber, M. J.:* In: Optical properties of ions in crystals (eds. *Crosswhite, H. M.,* and *Moos, H. W.*), pp. 467—484. New York: Wiley-Interscience Inc., 1967.

18. *Rahman, Hu.:* J. Phys. C., Solid State Phys. *5*, 306 (1972).
19. *Weber, M. J.:* Phys. Rev. *157*, 262 (1967).
20. *Reisfeld, R., Boehm, L., Ish-Shalom, M., Fischer, E.:* J. Phys. Chem. Glasses *15*, 76 (1974).
21. *Reisfeld, R., Hormodaly, J., Barnett, B.:* Chem. Phys. Letters *17*, 248 (1972).
22. *Reisfeld, R., Lieblich, N.:* J. Phys. Chem. Solids *34*, 1467 (1973).
23. *Jørgensen, C. K.:* Modern aspects of ligand field theory, Amsterdam: North-Holland Publishing Co. 1971.
24. *Jørgensen, C. K.:* Oxidation numbers and oxidation states. Berlin-Heidelberg-New York: Springer 1969.
25. *Reisfeld, R., Bornstein, A., Boehm, L.:* J. Solid State Chem. 1975, 14.
26. *Reisfeld, R., Bornstein, A., Boehm, L.:* J. Non-Crystalline Solids *17*, 158 (1975).
27. *Reisfeld, R., Eckstein, Y.:* J. Solid State Commun. *13*, 265 (1973).
28. *Reisfeld, R., Eckstein, Y.:* J. Solid State Commun. *13*, 741 (1973).
29. *Reisfeld, R., Eckstein, Y.:* J. Non-Crystalline Solids *15*, 125 (1974).
30. *Reisfeld, R., Velapoldi, R. A., Boehm, L.:* J. Phys. Chem. *76*, 1293 (1972).
31. *Weber, M. J.:* Phys. Rev. B *8*, 54 (1973).
32. *Reisfeld, R., Eckstein, Y.:* J. Chem. Phys. (1975).
33. *Wanmaker, W. L., Bril, A.:* Philips Res. Rept. *19*, 479 (1964).
34. *Reisfeld, R., Hormodaly, J.:* Unpublished Results.
35. *Weber, M. J., Matsinger, B. H., Donlan, V. L., Surratt, G. T.:* J. Chem. Phys. *57*, 562 (1972).
36. *Caspers, H. H., Rast, H. E., Fry, J. L.:* J. Chem. Phys. *53*, 3208 (1970).
37. *Weber, M. J., Varitimos, T. E., Matsinger, B. H.:* Phys. Rev. B. *8*, 47 (1973).
38. *Weber, M. J.:* Private Communication.
39. *DeShazer, L. G., Ranon, U., Reed, E. G.:* Spectroscopy of neodymium in laser glasses, USCEE Report 479, January 1974.
40. *Krupke, W. F.:* IEEE J. Quant. Electronics QE-10 *1974*, 450.
41. *Sorkies, P. H., Sandoe, J. N., Parke, S.:* Variation of Nd^{3+} cross-section for stimulated emission with glass composition. Brit. J. Appl. Phys., J. Phys. D *4*, 1642 (1971).
42. *Orbach, R.:* In Optical properties of ions in crystals (eds. *Crosswhite, H. M., Moos, H. W.*), p. 445. New York: Wiley-Interscience Inc. 1967.
43. *Miyakawa, T., Dexter, D. L.:* Phys. Rev. B *1*, 2961 (1970).
44. *Soules, T. F., Duke, C. B.:* Phys. Rev. B *3*, 262 (1971).
45. *Fong, F. K., Naberhuis, S. L., Miller, M. M.:* J. Chem. Phys. *56*, 4020 (1972).
46. *Kiel, A.:* Multiphonon spontaneous emission in paramagnetic crystals. In: Quantum electronics (eds. *Grivet, P., Bloembergen, N.*), pp. 765—777. New York: Columbia U. P. 1964.
47. *Riseberg, L. A., Moos, H. W.:* Phys. Rev. *174*, 429 (1968). — *Riseberg, L. A., Ganrud, W. B., Moos, H. W.:* Phys. Rev. *159*, 262 (1967).
48. *Englman, R., Jortner, J.:* Mol. Phys. *18*, 145 (1970).
49. *Reed, E. D., Moos, H. W.:* Phys. Rev. B *8*, 980 (1973).
50. *Reisfeld, R., Boehm, L., Eckstein, Y., Lieblich, N.:* J. Luminescence (1975).
51. *Reisfeld, R., Boehm, L.:* Unpublished Results.
52. *Nakazawa, E., Shionoya, S.:* J. Phys. Soc. Japan *28*, 1260 (1972).
53. *Reisfeld, R., Eckstein, Y.:* J. Non-Crystalline Solids *12*, 357 (1973).
54. *Flaherty, J. M., DiBartolo, B.:* J. Luminescence 8, 51 (1973).
55. *Kolobkov, V. P., Petrovskiy, G. T.:* Optical Technology *38*, 175 (1971).
56. *Nyquist, R. A., Kagel, R. O.:* Infrared spectra of inorganic compounds. New York-London: Academic Press 1971.

Received January 2, 1975

Structure and Bonding: Index Volume 1-22

C. K. JØRGENSEN

Oxidations Numbers and Oxidation States

VII, 291 pages. 1969
(Molekülverbindungen und
Koordinationsverbindungen in
Einzeldarstellungen)
Cloth DM 68,–; US $29.30
ISBN 3-540-04658-5

Contents: Formal Oxidation Numbers. Configurations in Atomic Spectroscopy. Characteristics of Transition Group Ions. Internal Transitions in Partly Filled Shells. Inter-Shell Transitions. Electron Transfer Spectra and Collectively Oxidized Ligands. Oxidation States in Metals and Black-Semi-Conductors. Closed-Shell Systems, Hydrides and Back-Bonding. Homopolar Bonds and Catenation. Quanticule Oxidation States. Taxological Quantum Chemistry.

Die Umsetzung von π-Elektronen-Systemen mit Metallhalogeniden führt in Anwesenheit von Protonen zu Proton-Additionskomplexen. Sind keine Protonen vorhanden, so entstehen π- oder σ-Komplexe. Besonders ausführlich werden die binären Systeme π-Elektronen-System und Metallhalogenid behandelt. – Während das Gebiet um die Jahrhundertwende nur von bescheidenem präparativen Interesse war, ist es heute ein eigenes faszinierendes Arbeitsgebiet geworden, in dem alle modernen spektroskopischen Methoden zur Aufklärung der Struktur- und Bindungsverhältnisse angewandt werden.

Prices are subject to change
without notice
Preisänderungen vorbehalten

H.-H. PERKAMPUS

**Wechselwirkungen von π-Elektronensystemen
mit Metallhalogeniden**

64 Abbildungen. 37 Tabellen
XI, 215 Seiten. 1973
(Molekülverbindungen und
Koordinationsverbindungen in
Einzeldarstellungen)
Gebunden DM 78,–; US $33.60
ISBN 3-540-06318-8

Inhaltsübersicht: Einleitung und Abgrenzung. Eigenschaften der Donatoren und Acceptoren. Proton-Additions-Komplexe. π-Komplexe. σ-Komplexe. Register.

Springer-Verlag
Berlin Heidelberg New York

Electrons in Fluids

The Nature of Metal-Ammonia
Solutions
Editors: J. Jortner, N. R. Kestner
271 figs. 59 tables
XII, 493 pages. 1973
Cloth DM 120,−; US $51.60
ISBN 3-540-06310-2

This full and up-to-date account of
the chemical and physical properties
of electrons in polar, nonpolar, and
dense fluids includes contributions
from both theoretical and experimental chemists and physicists,
thus clearly indicating the interdis-
ciplinary nature of this field.

Rare Earths

70 figs. III, 253 pages. 1973
(Structure and Bonding, Vol. 13)
Cloth DM 78,−; US $33.60
ISBN 3-540-06125-8

The crystallographic behaviour of
both mono- and polynuclear fluoro
complexes of uranium and trans-
uranium elements is thoroughly
discussed by Dr. Penneman and his
collaborators from the Los Alamos
Laboratory. Binary silicates, as well
as silicates containing additional
alkali ions, are exhaustively reviewed
by Dr. Felsche.
Vitreous silicates and borate, phosphate and germanate glasses are used
by Dr. Reisfeld as matrices for trivalent lanthanides showing narrow-
line and broad-band emission and
absorption spectra. The energy
transfer from one optically excited
ion to another is studied.
Dr. Jørgensen treats the connections
between the electron configurations
and the behaviour of the individual
lanthanides both in monatomic

entities and in compounds, the
refined spin-pairing energy theory,
and points out that the previously
insoluble problem of determining
4f ionization energies was solved in
1971 by photo-electron spectro-
metry of the solids.

H. RICKERT

Einführung in die Elektrochemie fester Stoffe

64 Abb. XV, 223 Seiten. 1973
Gebunden DM 46,−; US $19.80
ISBN 3-540-06266-1

Die Elektrochemie fester Stoffe ist
in den letzten Jahren weit über das
rein wissenschaftliche Interesse
hinaus gewachsen und hat an Bedeu-
tung und technischen Anwendungs-
möglichkeiten gewonnen. Wenige
Stichworte genügen, um das anzu-
deuten: Galvanische Ketten mit
festen Elektrolyten, Fehlordnungen
in festen Stoffen, Transportvor-
gänge, Festkörperreaktionen, Halb-
leiter. Chemiker, Physiker, Metal-
lurgen und Werkstoffwissenschaft-
ler sowie vor allem auch Studenten
finden hier Grundlagen und Bei-
spiele einer modernen Forschungs-
richtung verständlich dargestellt.

Prices are subject to change without notice
Preisänderungen vorbehalten

Springer-Verlag
Berlin Heidelberg New York